Biopolitics and the Philosophy of Death

ALSO AVAILABLE FROM BLOOMSBURY

The Politics and Pedagogy of Mourning, Timothy Secret
Dying for Ideas, Costica Bradatan
Desire in Ashes, Simon Morgan Wortham and Chiara Alfano

Biopolitics and the Philosophy of Death

PAOLO PALLADINO

Bloomsbury Academic
An imprint of Bloomsbury Publishing Plc

B L O O M S B U R Y
LONDON · OXFORD · NEW YORK · NEW DELHI · SYDNEY

Bloomsbury Academic
An imprint of Bloomsbury Publishing Plc

50 Bedford Square	1385 Broadway
London	New York
WC1B 3DP	NY 10018
UK	USA

www.bloomsbury.com

BLOOMSBURY and the Diana logo are trademarks of Bloomsbury Publishing Plc

First published 2016

© Paolo Palladino, 2016

Paolo Palladino has asserted his right under the Copyright, Designs and Patents Act, 1988, to be identified as Author of this work.

All rights reserved. No part of this publication may be reproduced or transmitted in any form or by any means, electronic or mechanical, including photocopying, recording, or any information storage or retrieval system, without prior permission in writing from the publishers.

No responsibility for loss caused to any individual or organization acting on or refraining from action as a result of the material in this publication can be accepted by Bloomsbury or the author.

British Library Cataloguing-in-Publication Data
A catalogue record for this book is available from the British Library.

ISBN: HB: 9781474283007
PB: 9781474282994
ePDF: 9781474283014
ePub: 9781474283021

Library of Congress Cataloging-in-Publication Data
A catalogue record for this book is available from the Library of Congress.

Typeset by Fakenham Prepress Solutions, Fakenham, Norfolk NR21 8NN

CONTENTS

Acknowledgements vi

   Introduction 1
1 Evental figures and questions of method 13
2 Ageing and the molecular way of life 37
3 The evolutionary biology of ageing and death 67
4 Molecularizing the biology of ageing and death 93
5 Forging the future 119
6 Life, death and philosophy 147
7 The arts of living and dying 173
   Conclusion 199

Notes 213
Works cited 243
Index 269

ACKNOWLEDGEMENTS

My debt to Mick Dillon, Tiago Moreira, Marcia Pointon, Thomas Söderqvist and Teresa Young is immeasurable. The argument advanced in *Biopolitics and the Philosophy of Death* is the product of ten and more years of discussion with them on how best to think about the task of critical reflection upon human embodied existence. Put simply, without their endless provocations, *Biopolitics and the Philosophy of Death* would never have taken form. I am just as indebted to Howard Caygill, Adrian Mackenzie, Rob Mitchell, Phillip Thurtle, and the manuscript's anonymous readers for helping to forge my responses to these provocations into something like a disciplined argument at the intersection of historical, philosophical and sociological modes of inquiry. As always, any remaining faults are entirely my own. My debt to Rob Kirk and Karen Rader for the moral support lent in the many moments of doubt about the merits of the entire project is equally great. Crossing disciplinary boundaries, without privileging one disciplinary mode over another, is a very testing enterprise. I would also like to stop and pause, to remember Paul Fletcher and John Pickstone, both of whom sadly passed away before these votes of thanks entered the public domain. Thank you all. Finally, *Biopolitics and the Philosophy of Death* is dedicated to my parents, who also passed away during these years, like two of the many other singular and irreplaceable older people who haunt these pages.

Introduction

> *Genealogy does not pretend to go back in time to restore an unbroken continuity that operates beyond the dispersion of forgotten things; its duty ... is to maintain passing events in their proper dispersion; it is to identify the accidents, the minute deviations – or conversely, the complete reversals – the errors, the false appraisals, and the faulty calculations that gave birth to those things that continue to exist and have value for us.*
>
> MICHEL FOUCAULT

How is one to make sense of the present moment? If the passage of time changes everything, transforming that which was once regarded as a matter of necessity into the accident of historical transformation, answering this otherwise simple question becomes a very complicated matter. The central conceit of *Biopolitics and the Philosophy of Death* is that to answer this question one needs to think about death, and to think about death not just as a philosophical concept, but also as a phenomenon of increasing interest to the biomedical sciences.

Reflecting upon the inevitability of ageing and death has a very long history that goes back to the founding moments of Judeo-Christian culture. According to Ecclesiastes, one of the books of the Old Testament, it seems inevitable that the world should grow poorer and darker the closer one approaches that moment when one is no more, so testing one's faith in the goodness of this world: 'Remember now thy Creator in the days of thy youth, while the evil days come not, nor the years draw nigh, when thou shalt say, I have no pleasure in them' (Eccl. 12.1). Martin Heidegger once sought to focus on just this experience to reconstruct the understanding of an authentic and truthful mode of existence in the aftermath of the 'death of God', this being Friedrich Nietzsche's memorable phrase for the modern loss of faith in the possibility of any transcendent moral compass and source of meaning. Situated between critical and historical modes of examination, *Biopolitics and the Philosophy of Death* will advance the proposition that, today, this notion that death will one day come to visit us all can no longer be regarded as so indubitable. As Aubrey de Grey (2005a), iconoclastic biomedical scientist, has suggested, 'the first person to live to be a thousand years old is certainly alive today; indeed, he or she may be about to turn sixty'. De Grey may well exaggerate, but there can be little doubt that recent discussions about the possibility of extending human longevity far beyond the traditional three score and ten years illustrate very powerfully how the course of human existence, from birth to death, has come to matter to the business of governing the contemporary civic polity and its constituent subjects, and to matter more than ever before.[1] At the same time, the task of critical reflection seems wholly at odds with this historically unprecedented situation. If the biopolitical terminology that is mobilized to understand such situations points to a novel mode of governance, which is organized today around life and its potential to proliferate beyond all limit, much of such reflection is underpinned by considerations about the very negation of life, namely death. This applies not just to Giorgio Agamben and his writings on the subject, which have done much to precipitate intense discussions about the place of death within the biopolitical ordering of the contemporary polity, but also to Michel Foucault and to all those other cultural critics who work in Foucault's wake. The challenge, yet others argue, is to construct an understanding of biopolitical existence that centres upon life, not death, and Gilles Deleuze is regarded increasingly as offering the required alternative perspective. *Biopolitics and the Philosophy of Death* focuses on the relationship between Foucauldian and Deleuzian modes of thought, especially as they touch upon the relationship between life and death.[2] Importantly, as Gil Anidjar (2011) has observed, contemporary discourse about of the meaning of life is fraught with difficulty.[3] Titian's 'The Three Ages of Man' conveys the nature of this difficulty much better than any words.

Most immediately, Titian's painting offers a figurative representation of the canonical, tripartite division of human, embodied

FIGURE 1 Titian (Tiziano Vecellio), 'The Three Ages of Man', about 1512–14 (NGL 068.46) Edinburgh, Scottish National Gallery (Bridgewater Loan, 1945)

existence by means of the trajectory from the sleeping *putti* on the one side of the frame to the two lovers on the other side, then to the old man in the distant background, who, holding the classic *memento mori* in his hand, contemplates his mortality. On the other hand, as Agamben has observed in *The Open* (2002), the one text where he comes closest to discussing the materiality of human, embodied existence, such a reading overlooks the dying or dead tree and how it constitutes a visual circuit, which returns from the old man back to the sleeping *putti*.[4] In so doing, Titian's painting could be said to counterpoise two very distinct meanings of life, namely life as the course of existence from birth to death, and life as the infinite movement of transformation. It is not clear whether and how these two meanings can be reconciled, except perhaps within something like the theological frame that Titian constructs by depicting a church, which is located in the far background and yet foregrounded by its casting against the horizon.

The thesis at the heart of *Biopolitics and the Philosophy of Death* is that any answer to the question about how one is to make sense of the present moment must confront this double meaning of life.

Provocations

The urge to juxtapose Michel Foucault, Gilles Deleuze and the contemporary development of biomedical understanding of ageing and death is rooted in a text that has proven very important to contemporary thought about how best to understand the changes wrought by the more general development of the biomedical sciences. In *Foucault* (1986), Deleuze's commemoration of Foucault's work, Deleuze maintained that Foucault had done much to advance critical understanding of the 'death of Man', the critical point when Man, understood here as the modern measure of all things, confronts the fundamental impossibility of understanding its own constitution. Deleuze also maintained, however, that Foucault had been unable to imagine what new formation might take the place of Man because his thought remained captive to a definition of life and death that privileged the organism. In other words, Foucault proved unable to forget the lessons Xavier Bichat imparted in his *Physiological Investigations of Life and Death* (1801). Deleuze preferred August Weismann's *Essays Upon Heredity and Kindred Biological Problems* (1891) and the very different understanding of the relationship between life, death and the organism found therein. Weismann, as will become evident, does indeed offer an understanding of this relationship that is much more attuned to the present situation and the attendant promises of immortality, but, as Elizabeth Grosz (2004) observes, Deleuze also read Weismann in a manner that seems less true to Weismann than it is to Henri Bergson. This difference is not without consequences for contemporary historical, philosophical and sociological thought. *Biopolitics and the Philosophy of Death* will examine these diverse readings and interpretations of biological understanding of the relationship between life, death and the organism, weighing them against the reconfiguration of ageing and death advanced by contemporary biogerontology, a relatively novel and sometimes controversial subfield of gerontological research that is engaged in the molecular and demographic characterization of the processes associated with the development of the organism from birth to death.[5] The aim of this double exercise is to test the limits of Deleuzian and neo-Deleuzian intimations that the reflective subject, like the mortal organism, must today give way to the constitutive dynamics of molecules and populations, to the play of life itself. There can be no doubting the productivity of current discussions about the importance of Deleuze to the development of a better understanding of the contemporary, increasingly complex relationship between the diverse practices that are involved in the movement across distinctions between human and non-human actors, as well as those between material and ideal structures.[6] There are, however, alternative ways of organizing such diversity, and the question *Biopolitics and the Philosophy of Death* seeks to pose ultimately is whether the accompanying rush to evacuate all networks of any constitutive centres and nodes is not sustained tacitly by the replacement of God and Man with Life Itself.[7]

Background

The critical enterprise at the heart of *Biopolitics and the Philosophy of Death* emerges from three related considerations.

First, the biogerontologists upon whom *Biopolitics and the Philosophy of Death* will focus equate ageing with the totality of the individual's life course, from cradle to grave, and then call not only for comprehensive health policy programmes to manage such totality, but also for the radical reorganization of biomedical research around such life course. This, in their view, is the only way to secure the health and longevity of the population. Furthermore, according to these biogerontologists' vision of how best to secure such healthier and longer lives, mundane practices like regular exercise and proper nutrition must be understood henceforth as supporting the molecular mechanisms that have emerged over the course of evolutionary history and that secure the integrity of the genome. As such, the biogerontological programme could be said to exemplify and to thus offer a useful site for critical reflection upon the more general developments that have led Nikolas Rose to posit the emergence of a new form of political subjectivity. In *The Politics of Life Itself* (2007), Rose proposes that contemporary subjects are constituted in the work of caring for their bodies and the exercise of their biological citizenship, on the understanding that life refers here to physiological processes that have a wholly contingent and very ambiguous relationship to the organism usually associated with the expression of these same processes. In other words, the politics of life itself would seem to signal the sundering of life and the organism.

Second, the very same biogerontological programme that could be said to exemplify the emerging politics of life itself is proving difficult to implement. This is partly because the programme rests on an understanding of death that August Weismann first articulated during the course of his fundamentally important investigations of heredity and evolution. On this understanding, death is to be regarded as a by-product of a division of biological labour under natural selection such that the mortal organism is a secondary product of the dynamics of molecules and populations. As Hans Jonas put it in *The Phenomenon of Life* (1966), 'in a reversal of the classical formula, one would have to say that the developed is for the sake of the undeveloped, the tree for the sake of the seed'.[8] Such reversal of relationships between constituent parts and object constituted, and the accompanying decentring of the organism, seems difficult to sustain. What the biogerontological programme would appear to seek instead is how best to integrate perspectives forged at three different biological scales, namely at the level of molecules, organisms and populations. Furthermore, such articulation would appear to involve some novel mode of organizing the institutions corresponding to each of these three levels, namely laboratories, clinics and communities. That this situation should be fraught with difficulty is not wholly surprising because, as Peter Keating and Alberto Cambrosio observe in *Biomedical Platforms* (2003), the biomedical enterprise has always sought to draw together the most disparate and incommensurable modes of thought and practice. Extending this understanding, *Biopolitics and the Philosophy of Death* will advance the proposition that the diversity that so characterizes the biogerontological programme can be ordered by observing how distinctive rhetorical tropes, such as the truth that death is inevitable and the hope that it can be postponed indefinitely, are deployed both repeatedly and in a patterned manner to articulate distinct forms of biopolitical existence. Furthermore, these rhetorical tropes can be organized according to equally distinctive orderings and aggregations of actors through which these same actors construct and adjust their positions in debates around the purpose and value of the biomedical enterprise. These formations, in other words, generate and perform distributions, defining or embodying a characteristic approach to what might, does, or should pass from whom to whom and under what circumstances. In so doing, these formations generate and perform distinctive distributions of power and agency. Finally, coordination must take the place of any hierarchical ordering of these formations. As *Biopolitics and the Philosophy of Death* will seek to demonstrate, coordinating the operation of the different actors that are involved in the articulation of the biogerontological programme requires thinking across the differences between what might also be called the anatomo-political, biopolitical and molecular regimes of truth. This is not easy, and it is not clear what the source of such difficulty might be, whether it be of ontic or ontological order.[9]

Third, the questions about life, death and the organism, which the biogerontological programme thus poses, seem important to understanding the relationship between Michel Foucault and Gilles Deleuze with regard to their shared ambition to reconstruct philosophy in a way that might overcome the long shadow of transcendental metaphysics, God's long shadow.

Foucault and Deleuze were equally concerned to advance Friedrich Nietzsche's attempt to evacuate philosophy of all transcendental attachments by turning to biology and to Charles Darwin in particular. At the same time, and in a sense wholly different to the other great critic of modern humanism, Jacques Derrida, both Foucault and Deleuze abjured any debt to Martin Heidegger, he who had tried to expurgate the conjunction of philosophical and biological thought of any understanding that the ever-proliferating heterogeneity of the phenomenal world might attest to some vital, immanent power, irreducible to either substance or extension. Foucault and Deleuze regarded Heidegger's emphasis on the mortality of the human animal as reaffirming the central importance of the reflective subject and its underlying metaphysical assumptions. If Heidegger maintained that the event of 'my death' was fundamentally important to the recasting of the philosophical project and that this event could not be rendered as an object of scientific inquiry, except by eliding all difference between my death and the death of others, it could only become intelligible by equating it with the experience of asymptotic movement toward the fateful and obscure passage into non-existence. In light of the seemingly inescapable, consequent presupposition of an experiencing subject, it is not surprising that Foucault should have then asked, in an interview toward the end of his life, 'rather than making a somewhat internal description of lived experience, shouldn't one, couldn't one instead analyze a number of collective and social experiences?'[10] Yet, as *Biopolitics and the Philosophy of Death* will seek to establish, Heidegger, Foucault and Deleuze's shared focus on embodied existence, and Foucault and Deleuze's simultaneous disavowal of Heidegger's powerful answer to the question about the future of philosophy, calls for closer attention to Foucault and Deleuze's own, arguably more positive understanding of death. In so doing, *Biopolitics and the Philosophy of Death* seeks to offer something like a counterpart to David Farrell Krell's *Daimon Life* (1992), a perceptive study of the conceptual links between Heidegger and the like of Eugen Korschelt, author of *The Duration of Lifespan, Ageing and Death* (1922). At the same time, however, *Biopolitics and the Philosophy of Death* also seeks to bring to bear upon this task Michel Serres' insights into the process of discursive clarification, especially as these insights have been reworked into a veritable sociology of critical capacities, in Luc Boltanski and Laurent Thévenot's *On Justification* (1991).

Boltanski and Thévenot offer not just a provocative understanding of the coexistence of multiple modes of reasoning and the consequences of such coexistence, an understanding that will inform the approach to the complexity of the biogerontological programme discussed above, but they also understand the social scientist's endeavour to clarify the nature of a social encounter as participating inescapably in the process described. Such reflexive considerations are crucially important to the genealogical approach that *Biopolitics and the Philosophy of Death* advances as the necessary prerequisite of any engaged, yet critical, response to the Deleuzian challenge. Strikingly, Nietzsche's development of the genealogical method was not wholly removed from his uneasy relationship to the Darwinian demotion of Man and God. Taken by the productivity of nature, which *On the Origin of Species* powerfully disclosed, Nietzsche understood the task of genealogical method in the following terms:

> For to translate humanity back into nature; to master the many vain and fanciful interpretations and secondary meanings which have been hitherto scribbled and daubed over that eternal basic

text homo natura; to confront man henceforth with man in the way in which, hardened by the discipline of science, man today confronts the rest of nature, with dauntless Oedipus eyes and stopped-up Odysseus ears, deaf to the siren songs of old metaphysical bird-catchers who have all too long been piping to him 'you are more! You are higher! You are of a different origin!' – that may be a strange and extravagant task but it is a task – who would deny that? Why did we choose it, this extravagant task? Or to ask the question differently: 'why knowledge at all?' – Everyone will ask us about that. And we, thus pressed, we who have asked ourselves the same question a hundred times, we have found and can find no better answer.[11]

At the same time, the thus denuded genealogist was no Darwinian human animal either. Nietzsche also understood that Darwinian notions of nature are far too intent on imposing rules and regulations, figments of the human imagination, and are thus untrue to the formative and exhilarating understanding of life as defying all constraint. Consequently, the genealogist who is properly attuned to the natural order of things as the domain of unfettered creativity must also be poised precariously on the verge of incoherence, if not on the verge of something altogether different and abyssal.[12] As Serres observed about the urge to *mimesis*, about the urge to be at one with the phenomenon described, and in a manner that resonates with the Derridean emphasis on the productivity of obscurities at the heart of language, 'to have light and clarity alone, one would have to inhabit the very source of light, or, I don't know, remove the medium and thus create a void. As soon as the medium intervenes, the ray of light energetically and haphazardly seeks its fortune. One sees because one sees badly. It works because it works badly.'[13] Bringing order to the phenomenal world is necessary to meaning and being, and so there is something deeply impossible about the genealogical injunction to resist the allure of unifying narratives and instead keep the passage of events 'in their proper dispersion', as Foucault once put it, in his own reflections on the genealogical method.

As such, the argument that *Biopolitics and the Philosophy of Death* seeks to advance in response to the question that opened these introductory reflections might be best regarded as unfolding between two images, between Titian's 'The Three Ages of Man' and a photograph of Antony Gormley's sculptural installation at Crosby Beach.

FIGURE 2 Antony Gormley, 'Another Place', 1997. Photograph: Rachel Docherty.

Gormley's strangely impassive, but nonetheless evocative figures recall the questions that Deleuze first posed in response to the famous closing lines of Foucault's *The Order of Things* (1966):

> As the archaeology of our thought easily shows, man is an invention of recent date. And one perhaps nearing its end. If those arrangements [of knowledge that sustain this invention] were to disappear as they appeared, in some event of which we can at the moment do no more than sense the possibility – without knowing either what its form will be or what it promises – were to cause them to crumble, as the ground of Classical thought did, at the end of the eighteenth century, then one can certainly wager that man would be erased, like a face drawn in sand at the edge of the sea.[14]

How is one to think and live in the present moment, abjuring God, Man and Life Itself? Posing this question is perhaps a first, necessary step toward the renewal of the attempt to free existence of the burden imposed by our Judeo-Christian culture.[15]

Outline of the argument

The argument that *Biopolitics and the Philosophy of Death* seeks to advance unfolds in three parts.

The first part of the argument, which is taken up in the first chapter, focuses on questions of method. Beginning with a complex, metaphorical representation of recent developments within biomedical understanding of ageing and death, the chapter draws attention to conceptual difficulties that neo-Darwinian and post-structuralist juridico-political theory would seem to share, and that are not unrelated to those involved in offering a properly critical account of the present moment.

Building on the conclusions drawn in the first part, the second of the three parts, which runs from the second to the fifth chapters, draws on Michel Foucault, seeking to articulate a rigorously genealogical account of the present moment. If one must make sense of the present moment by turning to biology because biology has today taken up the burden once shouldered by theology and humanism to speak about the contingencies of the human condition, one must start with the here and now of biological understanding of ageing and death. Consequently, the argument turns first to growing public concern about ageing, to the emergence of novel modes of biopolitical governance, and to the questions about these novel modes that the development of biogerontological approaches to understanding and managing ageing would seem to pose. There is no agreed understanding of ageing and death as objective, biological phenomena, only competing and yet complexly interrelated explanations. The subsequent two chapters outline two genealogical strands of the contemporary biogerontological moment, specifically the failed attempts to establish properly biodemographic and biomolecular perspectives on ageing and death. After having done so, the argument takes into consideration how contemporary biogerontology mobilizes this heritage to construct an alternative response to the challenges presented by an ageing population. Importantly, this second part comes to a close by observing how the implementation of the proposed integration of biomedical, biomolecular and biodemographic modes of thinking about ageing and death, as well as the institutional formations corresponding to these different modes, is proving difficult.

Taking advantage of a common point of reference, August Weismann's reflections on the biology of ageing and death, the third part of the argument, encompassing the sixth and seventh chapters, then examines how philosophical thought, when called to confront the mortality of the human subject, finds itself caught in the same difficulty as biogerontology. This third part does so by first considering how Gilles Deleuze responds to the questions about the fate of the human subject, which Foucault would seem to pose and leave unanswered, by proposing that the emergence of the human form is no more than a passing episode in the infinite movement of life itself. Despite the great differences that can then be said to exist between Foucault and Deleuze, the argument moves toward its conclusion by bringing their perspectives to bear on questions of life and death posed by so-called 'end-of-life decisions', those increasingly common and telling attempts to wrest control over the time and manner of the encounter with death. The proposition is that, in the last analysis, the mortality of the human form remains just as important to Deleuze as it is to Foucault and all those who work in Foucault's wake. *Biopolitics and the Philosophy of Death* closes by building upon this ambiguity, drawing on Luc Boltanski and Laurent Thévenot to propose a pragmatic answer to the difficulties confronting thought about diversity and difference, namely that the task of critical reflection may be better served by confronting how best to coordinate and integrate different ways of thinking about molecules, populations and the fate of the mortal organism, without transforming life itself into Life Itself, the new transcendent truth that would take the place once occupied by God and Man.

CHAPTER 1

Evental figures and questions of method

Time of Our Lives (1999), a popular account of technical developments within biogerontology, closes with an intriguing short story. The story begins with the words: 'After five days the Capsule had done its work, and Miranda lay dying.'[1] These words set the scene for a death-bed recollection in which Miranda, the dying narrator, weaves together autobiographical detail and a history of human society from the present day to her own present, which is some time after the middle of the twenty-second century. This historical narrative starts with nearly contemporary, disastrous attempts to extend human longevity beyond its present span, perhaps a tacit reference to the like of Aubrey de Grey and to their not wholly implausible promises of delivering human immortality. The grotesque results of such attempts, which recall the monsters in H. G. Wells's *The Island of Doctor Moreau* (1896), lead to the collapse of public faith in science and, eventually, to the collapse of the modern world order. A new order emerges from the ashes of destruction. By only allowing for demographic replacement, growth of the human population is much reduced and famine is ended. More importantly, while the ageing process is sufficiently well understood that regenerative medicine allows for effective immortality, those who choose to reproduce are required to forgo such immortality and so become a Timed One. Miranda is a Timed One. Miranda's death and its timing are determined by the artificial induction of biochemical processes leading to death at some random point between the visible onset of bodily degeneration and the physiological limits to the lifespan of reproducing organisms, but in a manner that will still allow Miranda to die with her family around her. Miranda's, in other words, is a 'good death'. The short story closes with one final, unanswered question:

> Lately, Miranda had learned, a new trend was appearing. People were choosing to make children earlier. And not only were first children being made earlier, but second ones too. The age at which people were becoming Timed Ones was growing younger. The trend was not yet statistically significant in view of the smallness of the sample, but it was definitely suggestive. If the trend was genuine, was it caused by a genetic or a psychological trend? ... 'Not my Special Project anymore,' mused Miranda. 'But I would have liked to know.'[2]

As a work of literary fiction, this short story is perhaps less than compelling, but there is nonetheless something very interesting about Miranda that deserves the closest attention, especially of anyone interested in a critical perspective on the genesis and conceptual implications of contemporary biomedical understanding of ageing and death.

As Thomas Kirkwood, author of *Time of Our Lives*, leads his readers to understand, Miranda's choice between immortality and reproduction is a metaphor for the balance operated by natural selection between investment in the continuity of the organism, its immortality being perfectly plausible, and the cost of such investment to the reproduction of the genetic line.[3] The normative implication is that we, the readers of *Time of Our Lives*, should avoid behaviours that undermine the continued viability of the mechanisms involved in repairing those molecular structures binding one generation to the next, and should engage instead in those activities that enhance the responsiveness of these repair mechanisms to the contingencies and vicissitudes of our lives. As Kirkwood puts it in the last lines of *Time of Our Lives*:

> Freedom makes us individually responsible for our choices and our actions. Is this why we so readily drug ourselves into inactivity with low-demand time-fillers when we could do so much? Let us be truly alive, so that when old age finally robs us of our vitality, we may feel that the time of our lives was well spent.[4]

In other words, much is staked on the transparency of the metaphor that is Miranda, the Timed One. At the same time, Miranda's last words also complicate the metaphor and in ways that Kirkwood may not have intended.

One cannot do without metaphors, Susan Sontag (1991) once observed, but they also are very deceptive.[5] Metaphors are mobilized to convey meaning where words, in themselves and of themselves, fall short of carrying the full weight of meaning, but they also erode all confidence of understanding whenever the reader hovers too long between the terms linked by the metaphor, to consider which term is clarifying which. In other words, transparency and univocity are not the attributes of metaphor. As Donna Haraway (1976) and others have observed, biology and medicine are not immune to such semantic obscurity and ambiguity.[6] Miranda herself is not just a figure for the spatially and temporally dispersed dynamics of evolution, but also a reflective subject who is very interested in understanding these same dynamics, elusive as they may be. Miranda, in other words, is at once the measure of all things and the product of invisible and determining processes beyond her understanding. There is something like a knot here, holding inside and outside together, and this so complex topological figure seems inseparable from Miranda's passing away, from her mortality. Miranda's wistful, last thought is: 'I would have liked to know.' Perhaps, however, the difficulty at issue is not so much one of knots to be untied, but closer instead to that presented by a 'Möbius Strip', the stuff of Maurits Escher's maddening figures and impossible geometries, geometries which will be equally familiar to the reader of the opening pages to *The Birth of the Clinic* (1964), where Michel Foucault equates the transformation of medical knowledge that was his subject with a transformation of spatial sensibility.[7] Whatever positive figuration one might choose, and the emphasis here is decidedly on the positive qualification, the central premise of *Biopolitics and the Philosophy of Death* is that Miranda, and all that for which Miranda is called to stand metaphorically, tests the limits of both neo-Darwinian biological thought and post-structuralist juridico-political theory, as the one seeks to decentre the organism and the other the subject.[8]

In sum, Miranda, this mortal, reflective figure is something of an evental figure, whose conflicting, positive attributes introduce a rupture in the order of things and a provocation to think differently about the fate of that other figure Foucault conjured in the closing pages of *The Order of Things* (1966), that figure that was destined to erasure, 'like a face drawn in sand at the edge of the sea'.[9] Over the course of the next two sections, this opening chapter of *Biopolitics and the Philosophy of Death* will first set out the issues that this disruptive figure raises for both biological and philosophical critiques, as well as draw attention to the distinctive place of August Weismann in the articulation of these critiques. The second of these two sections will instead outline the methodological difficulties involved in the attempt to examine the present moment, for which this same figure also stands, without privileging any particular method or set of presuppositions. In so doing, the outline will pave the way for the eventual turn to Luc Boltanski and Laurent Thévenot's *On Justification* (1991), to the scope it offers for the development of a pluralist perspective on the constitution of the present moment.

Critique between biology and philosophy

N. Katherine Hayles's and Rosi Braidotti's critical reflections on embodied existence are particularly instructive in regard to the many questions that Miranda would seem to pose.

Being interested in understanding in the relationship between science, technology and their configuration of embodied existence, Hayles is attracted to the conceptual structure of Richard Dawkins's *The Selfish Gene* (1976).[10] In this very important text, Dawkins wrote that:

> Individuals are not stable things, they are fleeting. Chromosomes too are shuffled into oblivion, like hands of cards soon after they are dealt. But the cards themselves survive the shuffling. The cards are the genes. The genes are not destroyed by crossing-over, they merely change partners and march on. Of course they march on. That is their business. They are the replicators and we are their survival machines. When we have served our purpose we are cast aside. But genes are denizens of geological time: genes are forever.[11]

Focusing on passages such as this, Hayles maintains that, for all its references to the temporally and spatially dispersed agency of natural selection such that organisms are of no great importance to the history of life, Dawkins and neo-Darwinian biological thought more generally struggle to dispense with all notion of a singular site of determination. Hayles writes:

> Dawkins's focus on the gene as the relevant actor radically decontextualizes genetic processes, abstracting the gene out of its embeddedness in the cell and indeed out of the organism as a whole. This decontextualization has the effect of reducing the relevant constraints, so that selective pressures are treated as if they operate on a single gene alone, not on interrelated groups of genes, genes located in cells, or cells operating within organisms. The *Selfish Gene* is underwritten by two imperatives: preserving the autonomous agency characteristic of the liberal subject, and re-locating it in the non-conscious modular units of the genes. These moves carry a double valence of anxiety and reassurance. Even though one's own agency is co-opted, agency itself is preserved. The key to the narrative is autonomy. To qualify as a 'real' actor in the drama, an agent has to be able to preserve its own identity and defend itself against encroaching foreign elements. The winners are those actors who can subvert and co-opt another's agency while keeping their own intact. In this sense Dawkins's gene is the ultimate individual, the triumphant product of that brand of Anglo-American ideology that ignores the complexities of social and economic contexts and declares success or failure to be solely the result of individual initiative.[12]

In other words, according to Hayles, Dawkins reinvents the *homunculus* conjured by early modern imagination, the fully formed and wholly autonomous little man within the maternal womb that served to explain the observed transmission of characters from one parental generation to the next, and, moreover, this *homunculus* greatly resembles the atomistic, rational maximizer of financial profit posited by neo-liberal economic theory. Not all biologists will agree with the broadest thrust of Hayles's criticism. There can be little doubt, however, that many of these biologists are exercised by the conceptual implications of the mode of explanation which Dawkins espouses, evoking as it does a regulatory site that is irreducibly simple and yet is at the very same time structured by complex temporal and spatial dynamics, and they are driven consequently to ask whether biology can really dispense with all 'need [of] an organism concept'.[13] Even if unclear what it is, there seems to be something irreducibly important about the organism.

As will become evident in later chapters, the organism and its role are very important to the evolution of contemporary biogerontological understanding of ageing and death. In the meantime, post-structuralist theory would appear to confront very similar difficulties as it seeks to decentre the modern, autonomous and reflective subject by positing that this subject is not primary and does not precede, but is instead the product of ontologically prior juridico-political processes. As Michel Foucault put it in the extended discussion of such processes, which he advanced most famously in *Discipline and Punish* (1975), modern culture may no longer regard the soul as the site of deliberation about right and wrongful action, but it would be a mistake to dismiss the concept altogether. He wrote:

> [i]t would be wrong to say that the soul is an illusion, or an ideological effect. On the contrary, it exists, it has a reality, it is produced permanently around, on, within the body by the functioning of a power that is exercised on those punished and, in a more general way, on those one supervises, trains and corrects, over madmen, children at home and at school, the colonised, over those who are stuck at a machine and supervised for the rest of their lives. This is the historical reality of this soul, which, unlike the soul represented by Christian theology, is not born in sin and subject to punishment, but is born rather out of methods of punishment, supervision and constraint.[14]

When viewed from this perspective, the subject is sometimes said to be akin to a textual construct, produced by the rules governing this subject's daily existence, from birth to death. As such, one is, down to one's very core and essence, the product of processes beyond one's understanding and agency. Hayles, the literary critic, seeks to further such understanding of the subject, but Hayles, the feminist critic, is snared by the long-standing difficulty concerning the relationship between discursive structures and embodied existence. When confronted with the full implications of the juridico-political constitution of the subject, namely the loss of all sense of embodiment and its formative importance, Hayles balks. As Arthur Kroker observes, in his aptly titled *Body Drift* (2011), Hayles is part of a select group of feminist theorists that includes Haraway and Judith Butler, and who are particularly concerned about how best to respond to the drift of post-humanist critical reflection away from the materiality of embodiment and the constraints it imposes upon the play of the symbolic order.[15] If some have responded to the difficulty by embracing Gilles Deleuze's endeavour to develop a fully materialist counterpart to the formative powers of the symbolic order, Hayles objects that this alternative would seem to reinstate the autonomous centre of deliberation posited by the modern, masculine imagination, albeit as it fantasizes about freedom from all forms of embodiment. Hayles then proposes another, altogether different response to the opposition of the humanist, autonomous subject and the post-humanist disembodied product of juridico-political processes. She writes:

> I prefer a third alternative, in which constraints act in dynamic conjunction with metaphoric language to articulate the rich possibilities of distributed cognitive systems that include human and nonhuman actors. Neither completely constrained nor entirely free, we act within these systems with partial agency amid local specificities that help to determine our behaviour, even as our behaviour also helps to configure the system. We are never only conscious subjects, for distributed cognition take place throughout the body as well as without; we are never only texts, for we exist as embodied entities in physical contexts too complex to be reduced to semiotic codes; and we never act with complete agency, just as we are never completely without agency. In a word, we are the kind of post-human I would want this word to mean.[16]

The more Hayles seeks to insist upon the relational character of the subject and yet retain some critical, formative role for embodied existence, the more she is thrown back toward some form of Cartesian 'cogito', that arguably most masculine figure who speaks the famous lines 'I think, therefore I am': 'In a word, we are the kind of post-human I would want this word to mean.' One need not agree with his explanation, but there would then seem to be some merit to Slavoj Žižek's observation that these critical enterprises struggle mightily to develop a coherent account of what might follow upon the evacuation of the modern subject. This subject, Žižek argues, is destined endlessly to return and disrupt all narratives of the end, and so much so that he prefaces his argument by writing that 'a spectre is haunting western academia ... the spectre of the Cartesian subject'.[17]

If Miranda, understood as an autonomous, reflective figure that is also the product of processes beyond her understanding, would appear to embody a conceptual difficulty common to neo-Darwinian biological thought and post-structuralist juridico-political theory, this is no accidental coincidence. Instead, the links that contemporary biogerontological understanding of ageing and death forges between the life course of the mortal organism, advances in molecular biology and growing understanding of the genetic structure of biological populations would appear to rest on the very same intellectual resources that have shaped philosophical thought about the constitution of the contemporary juridico-political subject. Braidotti's alternative understanding of the subject and its material embodiment helps to clarify the point.

Braidotti regards extant philosophical thought as resting upon historically and conceptually flawed foundations. Thus, reflecting upon the importance attached to the subject's mortality and finitude as the foundation for any discussion of life, Braidotti writes:

> From the position of an embodied and embedded female subject I find the metaphysics of finitude to be a myopic way of putting the question of the limits of what we call 'life' ... Death is overrated. The ultimate subtraction is after all only another phase in a generative process. Too bad that the relentless generative powers of death require the suppression of that which is the nearest and dearest to me, namely myself, my own vital being-there. For the narcissistic human subject, as psychoanalysis teaches us, it is unthinkable that 'life' should go on without my being there. The process of confronting the thinkability of a 'life' that may not have 'me' or any 'human' at the centre is actually a sobering and instructive process.[18]

Despite Georg Simmel's formative intuition that mortality inaugurates the very modern experience of individuality and subjectivity that is betrayed by Braidotti's authorial voice, the reflective process on which these words close paves the way 'for an ethical re-grounding of social participation and community building' and the renewal of 'hope'.[19] As Hayles might observe about this telling passage, it is unclear just how such expectation can be reconciled with the sundering of thought and embodied existence posited by the kind of post-humanism Braidotti espouses. Perhaps less enthralled by the promises of the so-called 'new materialism', which holds that there exists a material world independently of any knowledge of such world, and that what exists in such world and what is known about it are so constantly shaping one another as to defy all possibility of any stable point of reference, Hayles speaks more guardedly of a 'struggle to envision what will come after the fracturing of consciousness' and more equivocally about 'constructions of matter that matter for human meaning'.[20] More importantly, however, Braidotti's understanding of embodied existence rests on two pillars. It rests first on an explicit and uncompromising rejection of Martin Heidegger's famed insistence, in *Being and Time* (1927), on the fundamental importance of human mortality as the bedrock of existential awareness and autonomy, or, as Braidotti puts it, on the rejection of all presupposition of mortality 'as the trans-historical horizon for discussions of "life"'. Braidotti includes Foucault among those captive to such presupposition. Braidotti's argument also rests, secondly, on an equally explicit approbation of Deleuze's contrary assumption that, as Braidotti again puts it, 'the ultimate subtraction is … only another phase in a generative process'.[21] As *Biopolitics and the Philosophy of Death* seeks to establish, the two perspectives on death that Braidotti thus mobilizes, Heidegger's and Deleuze's, are indebted to different readings of the very same source, namely August Weismann's formative reflections on the biology of life and death.[22]

Weismann, founding figure of modern biology, is sometimes hailed as 'one of the great biologists of all time', and he is so important that both Henri Bergson and Sigmund Freud felt compelled to respond to his reflections on life, death and the organism.[23] Dawkins himself wrote that his ideas about genes and the role of the organism in evolutionary history were 'foreshadowed by A. Weismann in pre-gene days at the turn of the century'.[24]

Admittedly, the difficulties that this heritage presents for Deleuzian thought are considerable, and these difficulties have led some critics to minimize, if not ignore, Deleuze's debt to Weismann and Weismannism. These critics emphasize instead developmental perspectives on the evolution of biological structures, referring to Albert Dalcq and Raymond Ruyer in particular, and moving from there to contemporary criticism of 'molecular Weismannism'.[25] Yet, if developments within biology are as important as these endeavours would seem to suggest, the challenge within biology is not to replace Weismann and Weismannism, as much as to find a way of reconciling the different approaches to explaining the diversity of biological phenomena, of reconciling the different ways of thinking about molecules, populations and the mortal organism.

In sum, Weismann is not just a critically important figure in the evolution contemporary thought about the biology of ageing and death, but is just as important to the evolution of post-structuralist understanding of what it is to be human and to be invested in understanding the truth of the human condition. The challenge *Biopolitics and the Philosophy of Death* then seeks to outline is how to make sense of the complexities of the present moment in a pluralist manner that will acknowledge the mutual entanglement of biology and philosophy, and that will do so without collapsing one into the other because neither can claim any hold on certainty, even the certainty about the importance of human mortality. The remainder of this chapter is dedicated to the examination of the methodological difficulties involved in confronting such a challenge.

History, genealogy and historical criticism

The complications involved in any attempt to examine the resonance between different modes of thought about the relationship between life, death and human existence are many, especially if there is to be no privileged standpoint. Reflection on the historical genesis of this fraught relationship is equally caught up in the difficulty, but in a manner that perhaps discloses even more incisively the stakes at the heart of *Biopolitics and the Philosophy of Death*.

One cannot but begin with Friedrich Nietzsche, the first philosopher to confront both the importance of biology to modern thought and all the complications arising from such importance.

Writing in the long shadow of *On the Origin of Species* (1859), Nietzsche insisted that, among other things, any properly critical reflection upon the growing importance of biology to the development of modern thought required the adoption of a critical approach to the analysis of received historical narratives.[26] Such an approach entailed ultimately the abandonment of history for genealogy because only this latter mode could bring historical consciousness into 'the service of life'.[27] The differences between the two modes of historical explanation are often overlooked, so it is important to examine more closely Nietzsche's distinction. In *On the Genealogy of Morals* (1887), Nietzsche sought to criticize contemporary morality by proposing that it had developed into its existing form as a result of evolving political relations and not as the result of any dispassionate reflection upon the relationship between right and wrong. Famously, he opened such criticism by voicing scepticism about contemporary philosophers' claims with regard to the evolution of modern moral values because these philosophers had interpreted the entire history of morality in terms of utility, claiming first that the 'good' and the 'useful' were once one and the same, and secondly that such truth had been forgotten after centuries of habituation to thinking otherwise. From Nietzsche's perspective, not only did this understanding conflate the self-regarding beliefs of the aristocratic few, those who were used to assuming that what served their interests was just, useful and good, and the resentful and incommensurable insistence of the demotic multitude that the useful and the good were not one and the same, but it also overlooked along the way the violent struggles between these two parties. Furthermore, such mistaken understanding rested on these philosophers' habit of viewing the past through the lens formed by the moral biases of their own time and place. Importantly, while Nietzsche did not regard such biases as peculiar to the utilitarian moral philosophers he was addressing, he sought to embrace, rather than to foreclose, the conflict between the philosophical and historical modes of explanation that he thus disclosed. If it was to free human creative potential, historical narrative, Nietzsche argued, had to work from the present and not from the past to the present, seeking to unearth in the process some necessary, causal logic running from originating event to contemporary effect. Genealogical critique sought to operate in this preferred manner.

The difference that Nietzsche drew between historical narrative and genealogical criticism has proven fundamentally important to the evolution of Michel Foucault's thought about truth and the embodied subject. While Thomas Flynn (2005), acute and insightful commentator upon Foucault's historiographical method, regards the rupture between Foucault's archaeological and genealogical approaches as overstated, there can be little doubt that Foucault first turned his attention to genealogical method in the wake of critical response to *History of Madness* (1961), *Birth of the Clinic* (1964) and *The Order of Things* (1966). Though sharing Nietzsche's criticism of reason and morality, Foucault was sceptical about the heroic construction of the genealogist that seemed to emerge from Nietzsche's writings on the matter, and he sought consequently to revise the genealogical approach, aiming to construct a form of historical narrative that was more attentive to the position of the subject who traces the historical development of people and societies. Foucault's genealogy of the subject would account for the constitution of knowledge, discourses and the objects they conjure, striving to do so without having to refer to a subject who was either transcendent in relation to the field of events or who persisted unchanged throughout the course of historical transformation. Thus, in 'Nietzsche, Genealogy, History' (1971), Foucault explained how his ideas about genealogy were greatly influenced by Nietzsche's thought on the historical development of morality, but he also insisted that genealogy should display the plural and sometimes contradictory versions of the past involved in the formation of any received truth and certainty, including the truth of the historiographical self. Foucault wrote:

> Genealogy does not pretend to go back in time to restore an unbroken continuity that operates beyond the dispersion of forgotten things; its duty ... is to maintain passing events in their proper dispersion; it is to identify the accidents, the minute deviations – or conversely, the complete reversals – the errors, the false appraisals, and the faulty calculations that gave birth to those things that continue to exist and have value for us.[28]

These are powerful words and there can be little doubt that the allure of lending meaning to the world by beginning with 'those things that ... exist and have value for us' is great, but this historiographical approach also comes at a cost. This cost is hidden in Foucault's further insistence that the truth of which he spoke included the narrating subject, understood as a fully embodied being, because 'nothing in man – not even his body – is sufficiently stable to serve as a basis for self-recognition or for understanding other men'.[29] As Giorgio Agamben observes in his reflections on Foucault's historiographical method, not only is the historian's identity implicated in the act of historical reconstruction, but the genealogical method must also redouble the difficulty because its 'generative grammar' is inseparable from the historical transformation of embodied existence that is today anchored in the 'genetic code of *Homo sapiens*'.[30] Replaying the issues that Miranda raises, one is thus driven to ask what is the nature of that point of inflection, that turning point, that is 'us'. The prospect of an impossible conceptual structure returns to the fore. If Foucault, the parrhesiast who sought to speak the truth as clearly and bluntly as is possible, was all too aware of how entangled this complex situation was with the history of the Judeo-Christian culture to which he and much of contemporary culture are heirs, it is worth clarifying further what is the alternative understanding of historical consciousness which he, and Nietzsche before him, sought to displace by turning from history to genealogy.[31] For reasons that will become clearer as the argument proceeds and that are related to Agamben's criticism of Foucauldian thought, from *Homo Sacer* (1996) onward, the work of clarifying the historiographical stakes involved is best undertaken by turning to Jacob Taubes, other astute reader of Nietzsche's *On the Genealogy of Morals*.[32]

Taubes opened *Occidental Eschatology* (1947), his much overlooked discussion of historical process, by writing:

> The course of history is borne away in time ... The nature of time is summed up by its irreversible unidirectionality. From a geometrical point of view, time runs in a straight line in one direction. The direction of this straight line is irreversible.[33]

The meaning of this trajectory, Taubes argued, is inseparable from the 'end of days'. According to Taubes, Gnosticism is the central problematic of Judaic and Christian theologies. Their shared monotheism posits the unity and meaningfulness of the diversity and multiplicity that characterizes the phenomenal world, the world of all things apparent. This, however, requires some explanation with respect to the reason why such unity and meaning are not self-evident. Not only must the community of believers and their God then be alienated from one another, but the realization of their unity must also take the form of a revelation deferred. Furthermore, such revelation must disclose the falsity of the phenomenal world and, as such, the realization of the truth revealed requires the overturning of the existing order of the world. Judeo-Christian theology, in sum, cannot then but be eschatological and apocalyptic, and this essence, Taubes maintained, is disclosed by the millenarian tendencies of medieval Christianity. The enduring importance of this onto-theological framework and its structural implications, Taubes also maintained, lie in the transformation operated by the 'Copernican turn'. This, in Taubes's vocabulary, is both a literal, historical statement and a cipher, borrowed from Immanuel Kant, for any epistemological shift comparable to that from the closed *cosmos* of the medieval imagination, a *cosmos* divided between the eternal recurrence of the movements of the heavenly spheres and the rule of time within the human, earthly sphere, to the modern presupposition of a decentred, infinite universe. In the course of this transformation, Taubes observes, theological thought turned inward, to spirit, metaphysics was displaced by historical process, and the historical process itself became the vehicle of spiritual redemption. Taubes also argues that this reconstruction of the onto-theological framework in a manner adequate to the death of God, that is, to the loss of all possibility of direct communion with the heavens, is indebted to G. W. F. Hegel, Søren Kierkegaard and Karl Marx. In their very different ways, these three understood historical development as a matter of revolutionary struggles to break human action free of all necessity and so redeem all evil. Viewed from this perspective, that one was an idealist and the other two might be regarded as materialists, or that one of these two emphasized psychological states and the other economic relations, matters little. Taubes thus brought *Occidental Eschatology* to a close by writing:

> With Hegel on the one hand and Marx and Kierkegaard on the other, this study is not simply closed but essentially resolved. For the entire span of Western existence is inscribed in the conflict between the higher (Hegel) and the lower (Marx and

Kierkegaard) realms, in the rift between inside (Kierkegaard) and outside (Marx).[34]

However indebted to Nietzsche's *On the Genealogy of Morals*, this perspective on the diversity of modern thought is very far from either Nietzsche's or Foucault's understanding of the historiographical enterprise and its relationship to the freedom of the present moment.

In light of these considerations it is perhaps unsurprising that Taubes and Foucault should appear to have held very little in common. In fact, Paul Rabinow, when first examining what intellectual resources might be most useful to understand the transformations wrought by the contemporary development of the biomedical sciences, seeks to draw out the profound differences between Foucault's understanding of historical transformation and important contemporaneous debates over the nature of historical process, particularly the debates over the 'legitimacy of the modern age'. As Rabinow observes, Hans Blumenberg, while reflecting on Max Weber's thesis about modernity, reason and the process of secularization, insisted upon the novelty of modern culture and that it should not be understood as derivative, as shaped by past questions. Taubes was an active participant in the debate that Blumenberg thus precipitated and, as the preceding paragraphs illustrate, he clearly thought very differently to either Blumenberg or Rabinow.[35] Yet, for all the proximity between Foucault and Blumenberg, which Rabinow thus constructs, and, conversely, for all the distance between Foucault and Taubes, it is also the case that Taubes and Foucault shared great concern about the fate of the human figure within the modern age, and that they reflected upon the very same resources in the course of articulating such concern.[36] The consequences of such complexity are important to the argument that *Biopolitics and the Philosophy of Death* seeks to advance.

Wholly unpromisingly, Taubes's argument about the nature of historical process rested on the sharpest division between nature and culture, between that which is given and that which is made. As he puts it in *Occidental Eschatology*,

> The essence of history is freedom ... Freedom alone lifts mankind out of the cycle of nature into the realm of history. To exist in freedom is the only way that mankind becomes part of history.

Nature, and mankind embedded in nature, have no history. Freedom, however, can only reveal itself in apostasy. For as long as freedom is caught up in the divine cycle of Nature, it is subject to the necessity of God and Nature. A *non posse peccare* is no different from a compulsion to do good. Only mankind's answer to the word of God, which is essentially a negative one, is evidence of human freedom. Therefore, the freedom of negation is the foundation of history.[37]

The separation between nature and culture, according to Taubes, is the essential prerequisite for any true expression of human freedom. As he explains, such freedom began to take shape when humanity broke with nature and all aboriginal myth, as well as with classical culture, or, in other words, humanity only emerged into its own once it refused to be bound by the requirement of necessity and law. To 'the static ontology of Hellenic, Hellenistic philosophy [sic]', Judeo-Christian monotheism counterpoised a divine order wherein time was installed as unidirectional and irreversible movement from origin to end, at which final point the apostate's actions would be judged.[38] Importantly, in the course of the transformation of eschatological thought wrought by Christianity, beginning particularly with Origen and Augustine, the human subject's awareness of their mortality, the life course from birth to death, became critically important to the articulation of this historical trajectory, but in a sense wholly different to that which might be said to have obtained within aboriginal myth. Because it becomes a point of passage in a complex set of further transformations, death, on this later understanding, has nothing to do with death in the aboriginal 'cycle of birth and death'. Thus, when Taubes writes that 'time emerges when the eternity of the origin is lost and the order of the world is gripped by death', and immediately thereafter that the 'entanglements and disentanglements of death and life take place in history', the terms life and death refer to forms of being that must exceed the vicissitudes of embodied existence.[39] As such, Taubes could be said to contribute to understanding the genesis of the question Gil Anidjar poses when he asks how modern culture became biocentric, that is, how it came to posit embodied existence as absolutely central to all understanding and how it did so in a way such that life came to be regarded as no more than mundane species existence, and, at the same time, also absolutely sacred. Anidjar proposes that this situation is inseparable from the history of Christianity. As he puts it, quoting Hannah Arendt, 'only when the immortality of individual life became a central creed of Western mankind, that is, only with the rise of Christianity, did life on earth also become the highest good of man'.[40] While such consecration of life sets the theme running through *Biopolitics and the Philosophy of Death*, Taubes's deep suspicion of any naturalistic explanation of human, embodied existence is evident in his essay on Marx, Weber and Arnold Gehlen. After introducing the notion of the 'fetishization of the commodity', which he understands as the hidden intervention of social institutions whereby 'living work' is transformed into a commodity and the 'creator becomes a creature', Taubes proceeds to examine from this perspective both Weber's notion of the human subject's fateful subordination to the forces of rational organization and, most importantly, the naturalization of this process in Gehlen's sociology of institutions. Quoting Gehlen, Taubes writes that:

> The question about the relation of technological emancipation to society for us today turns into a question about the relation of technological emancipation to human emancipation. If it is said of technology that it is to be understood 'less as a human effort toward the propagation of material power, and more as a biological process on a grand scale', that is, as a meta-human process, which as such eludes human control, then this is fetishization in the most precise sense of the word.[41]

Reflecting further upon contemporary technologies of extermination and annihilation, and once again quoting Gehlen, Taubes then concludes that:

> Because technological emancipation could take place while human were incinerated and consumed by their own creations, because technological emancipation could take place while the institutions of industrial society devour them according to the laws of a Moloch ritual, to which men surrender knowingly, then, as a result, technological emancipation is not human emancipation. We are still very far from exiting history to arrive in *post-histoire*. Rather, we still stand deep in prehistory, where the sacrificial ritual of institutions is celebrated.[42]

The integration of technology and humanity, the latter in its existence as the human animal, does not lead to greater freedom, but to destruction and death. By regarding this as a tragedy, Taubes argues, Weber and Gehlen are not so much guilty of fetishism, as they are of complicity with the destruction and death wrought. They wilfully mistake the historically specific situation for the ontological foundation of human existence.[43]

Strikingly, the very same Gehlen appears to have been one of the sources of inspiration for Foucault's own reflections on the possibility of constructing a truly meaningful and authentic life within the modern age. During the lectures at the Collège de France on *The Courage of the Truth* (1984), Foucault turns to Cynicism and its legacies. After bemoaning the paucity and poverty of sources on the history of Cynicism, Foucault writes:

> In these interpretations, whether of Gehlen, Heinrich, or Tillich, Cynicism is always presented as a sort of individualism, of self-assertion, an intensification of the specific existence, of natural and animal existence, of existence at any rate in its extreme singularity, whether this is in opposition or reaction to the break up of the social structures of Antiquity, or faced with the absurdity of the modern world. By basing the analysis of Cynicism on this theme of individualism, however, we are in danger of missing what from my point of view is one [of its] fundamental dimensions, that is to say, the problem, which is at the core of Cynicism, of establishing a relationship between forms of existence and manifestation of the truth. It seems to me that it is the form of existence as living scandal of the truth that is at the heart of Cynicism, at least as much as the famous individualism we are in the habit of finding so frequently with regard to everything and anything. Well, if we were to agree – these are hypotheses, for possible work – to consider the long history of Cynicism on the basis of this theme of life as scandal of the truth, or of style of life as site of emergence of the truth (*bios* as alethurgy), it seems to me that there are some things we could bring out and tracks we could follow.[44]

Gehlen, in other words, may be mistaken because he presumes individualism, but what is important is that he contributes nonetheless to the recovery of a mode of being that is absolutely denuded of all pretence about the nature of the human animal and that is committed to the endless work of putting to the test every single assumption about both its life and the meaning of such life. The genealogical method would seem to participate in this understanding of critique and its future. As Foucault puts it in the closing lines of his essay on Nietzsche, genealogy and history, the function of 'critical history' is to risk 'the destruction of the subject who seeks knowledge in the endless deployment of the will to knowledge'.[45]

Viewed against this complex background, there would appear to be a conceptual knot at the very heart of *Biopolitics and the Philosophy of Death*. If testing all limits and self-understanding of one's subjectivity is the deeper meaning of the genealogical method, this same understanding of historiographical practice suggests that to think about how to write a history of ageing and death today must prove far from easy. If the site of embodied existence must be regarded as the site from which truthful reflection upon the subject's existential condition emerges and the terms of such existence are no longer as indubitable as they once seemed, death no longer being what it once was, the methods of reflection upon the present moment seem to be as deeply entangled as Agamben maintains in the very historical situation to be explicated. The alternative is, of course, to regard this knot as the product of festishism, having mistaken the historically specific situation for the ontological foundation of human existence, or to seek instead a new plane of consistency wherein all such knots are dissolved, but this latter gesture would seem to participate in a further turn of the onto-theological machine.[46] From this perspective, coming to terms with all that for which Miranda stands metaphorically begins to assume considerable importance, but this knot is important not just to the historian and the philosopher. It is also important to the sociologist.

Entangled regimes of knowledge

If *Biopolitics and the Philosophy of Death* seeks to make sense of the present moment by drawing out some of the questions that contemporary understanding of ageing and death would appear to pose, and it is also intent on remaining true to the complexity of the present moment by seeking to keep events and modes of thought in their proper dispersion, there then is something very alluring about Luc Boltanski and Laurent Thévenot's *On Justification* (1991).[47]

Boltanski and Thévenot are renowned for their role in a renewal of pragmatism within sociology whereby the actors observed are no longer regarded as calculating, self-interested, and therefore wholly transparent agents, and the social scientist himself or herself must also be regarded as being implicated in the production of social reality.[48] Historically, Boltanski and Thévenot's distinctive contribution to such renewal emerges from the combination of Boltanski's interest in the relationship between human action and the structures of political economy with Thévenot's formative insistence that, to understand economic activity, it is necessary to attend to the conventions that ensure the very possibility of economic exchange. Importantly, Thévenot suggested further that these conventions are both cognitive and moral insofar as they not only support mutual understanding between economic agents, but they also articulate particular qualifications of the objects and persons involved. In *On Justification*, Boltanski and Thévenot draw these strands together to pose a very basic question about how any two persons, for all the differences between them, come to those modes of coexistence manifest everywhere one might turn. They cannot give any precise answer to such a long-standing question, which goes all the way back to Plato and Aristotle, but, while articulating the question's fullest implications, they develop a usefully complex understanding of human action and the process of critique. They do so by rearticulating the traditions of functionalist sociology to argue that actors' engagement and attempts to harmonize their actions with other actors are linked intimately to the cognitive framing of objects and persons. In other words, in those situations where there is both dispute and an attempt to find some resolution, actors aim to render their different understanding of the situation comparable, but in so doing, these actors cannot refer to the situation at hand and must instead build arguments around some notion of the 'common good'. Crucially, Boltanski and Thévenot regard actors' accounts of their situation as irreducibly constitutive of the social processes described, and they also understand actors' justifications as requiring investment and durable commitment to one mode or another of understanding their situation. From this perspective, actors' articulation of the common good entails a constitutive projection beyond the situation at hand, and it is constrained by local weighing of the costs associated with drawing on some established moral order, as opposed to generating a new one. Consequently, Boltanski and Thévenot also come to regard the frames of reference involved as potentially innumerable. In *On Justification*, they draw on their empirical findings to offer six different modes of constructing the common good, which they associate with the names Augustine, Jacques-Bénigne Bossuet, Thomas Hobbes, Jean-Jacques Rousseau, Adam Smith and Henri de Saint-Simon. With each name goes a distinctive way of ordering things, standards and metrics, or, put most simply, distinctive ways of arranging life with others. As Boltanski and Thévenot put it in their prefatory remarks:

> Readers of this book may find it somewhat discomfiting not to encounter a familiar cast of characters: none of the groups – social classes, blue-collar workers, white-collar workers, youth, women, voters, and so on – with which we have become acquainted thanks to the social sciences and the quantitative sociological data that proliferate today; none of the 'men without qualities' whom economists call 'individuals' and who serve to buttress analyses of rational choices and preferences. Nor are there any of the life-size characters who have been appropriated for the realm of scientific knowledge by sociology, history, or anthropology in their most literary forms, much the way similar figures are highlighted by journalists or novelists. Short on groups, individuals, and persons, our book nevertheless abounds in beings, some of them human, some of them things. Whenever these beings appear, the state in which they operate is always qualified at the same time. The relation between these person-states and thing-states (which constitutes what we define as a situation) is the object of our study.[49]

While *Biopolitics and the Philosophy of Death* adopts this same mode of representation, Alex Comfort, Leonard Hayflick and Thomas Kirkwood taking the place of Boltanski and Thévenot's six figures, what is important about what Boltanski and Thévenot do with the relations between their 'person-states and thing-states' is, first, that the different orders of morality that they thus construct coexist with one another. In other words, these orders can equally be invoked by anybody in order to criticize, to justify or to reach a compromise with someone else. Any enterprise of understanding the constitution of that which is in common must come to terms with the extraordinary diversity of things that people do with words and things, and how arguments sometimes work in unison and sometimes not, so resulting in uncertainty, conflict and strife. Secondly, such necessarily unstable, but nonetheless generative, engagement between different orders of reason is driven by the very urge to constitute the common good and by the accompanying call to justify one comportment in relation to others. In other words, the very process of clarification and enlightenment, which is necessary to construct the social bonds sought, is the source of instability. As Boltanski and Thévenot write:

> The presupposition of a common good is required in order to establish a compromise. But the compromise will not hold up if the parties involved try to move ahead toward clarification, since there is no higher-ranking polity in which the incompatible worlds associated in the compromise can converge. An attempt to stabilize a compromise by giving it a solid foundation thus tends to have the opposite effect. An effort to define the common good that is supposed to sustain a compromise may actually shatter the compromise and shift it back into discord.[50]

As Boltanski and Thévenot themselves must admit, it is not clear how their consequently more complex understanding of the polity and its life is to be translated into norms of conduct, except at the cost of contradicting the formative insight that the social scientist must be regarded as being implicated necessarily in the production of social reality by either mobilizing one of the orders posited or by forging some novel one.

Biopolitics and the Philosophy of Death draws on Boltanski and Thévenot's insight into the full complexity of contemporary existence. The aim is to advance the notion that the contemporary subject, that knotted figure to which Miranda speaks very powerfully and whose disclosure seems inseparable from the moment of confrontation with mortality and finitude, should be understood as produced at the intersection of incommensurable and irreducibly different formations, in the contemplation of the impossibility of any supervening, rational and analytical ordering. God is dead, and so is Man. If some will look to Life Itself as a new ordering principle, it is not clear how this changes anything and what is to be gained.

CHAPTER 2

Ageing and the molecular way of life

Having sketched the conceptual and methodological difficulties involved in articulating an understanding of the present moment that is attuned to the relationship between, on the one hand, contemporary understanding of the biology of ageing and death, and, on the other hand, the complexities of historical, philosophical and sociological thought, this next chapter of *Biopolitics and the Philosophy of Death* inaugurates the second part of the argument advanced. This second part is dedicated to the more systematic, historically situated exploration of all these difficulties. As such, this particular chapter sets out to examine more closely the challenges confronting contemporary biomedical understanding and management of an expanding fraction of the population, the elderly.

After some brief and preliminary considerations about the periodization of biomedical understanding of ageing and its historical development, the chapter will turn its attention to reports issued in the United States by the Alliance for Aging Research, and in the United Kingdom by the Mass Observation Research Institute. It would seem that the ambivalence of public responses to contemporary research into the biology of ageing, which these reports disclose, may be due, at least in part, to disagreement among biomedical researchers about the aims of such research, whether it should aim to transform fundamentally the human condition, or whether it should aim instead to extend the great gains in longevity achieved during the past century and a half. The chapter will then turn to a transatlantic, biogerontological critique of contemporary biomedical research as beholden to outmoded ways of understanding the relationship between health and illness and to equally outmoded ways of securing the health and longevity of the population. Against this background, the chapter will introduce Paul Rabinow and Nikolas Rose's notion of an emerging biosocial subjectivity. Importantly, the discussion will focus on the manner in which the practical proposals emerging from the transatlantic, biogerontological critique of biomedicine would seem to exemplify this new biosocial subjectivity. In so doing, this discussion will also draw out how these same practical proposals rest on a neo-Darwinian understanding of the organism that disrupts pivotal distinctions between the mode of biosociality and the sociobiological reduction of social existence to a matter of molecular components and the dynamics of populations. The disjunction highlights the nature of the challenge confronting both biogerontology and contemporary social theory, namely how best to reconcile, on the one hand, the dynamics of molecules and populations constituting the mortal organism and, on the other hand, the human subject's opposing desire for a healthier and ever longer life. This exercise will prepare the ground for both a genealogy of the contemporary biogerontological moment, which will unfold over the ensuing two chapters, and for the answers biogerontology has sought to forge by drawing upon this heritage.

Ageing, death and governance

As Kathleen Davis (2008) observes, periodization, the drawing of boundaries between past, present and future, is a most difficult and politically charged enterprise. Thus, almost four decades ago, Peter Medawar, the recipient of the Nobel Prize for Physiology and Medicine for his contributions to immunology, who was also renowned for his biomathematical studies of the genetics of ageing, complained that 'the great public and private agencies are not competing with each other in their endeavours to support research on ageing'.[1] There was something deeply paradoxical about Medawar's complaint. By complaining as he did, Medawar was marking the end of a period of great scientific interest in ageing which, as Andrew Achenbaum (1995) and Hyung Wook Park (2008, 2012) have intimated, can be said to have started in the 1930s, in the midst of a global economic depression and the endeavours to mitigate the consequent strains upon the fabric of society. The intervening period had witnessed the proliferation of theories about the nature of the phenomenon within the biological and social sciences, but, over time, they had failed to garner much political support for the nascent gerontological enterprise. At the very same time, however, new developments were revitalizing gerontological research, beginning with both the establishment of the United States National Institute of Aging and the accompanying constitution of Alzheimer's Disease as the defining disease of old age, and the simultaneous, accelerating convergence of research on the molecular biology of cancerous cells and the regulation of cellular replication within the National Cancer Institute, in the United States and the National Institute of Medical Research, in the United Kingdom. By the 1990s, scientific interest in the biology of ageing was well-established, fuelled by the increasing integration of developments in molecular biology and new understanding of the demographic dynamics of longevity forged within the new field of biogerontology, so offering plenty of material for the production of Stephen Hall's *Merchants of Immortality* (2003), a popular account of such developments and the accompanying propositions about the possibility of greatly extending human longevity. If, in 1952, Medawar (1952) had identified ageing as a major problem confronting biology and medicine, in recent years biogerontologists such as Robin Holliday (2006) and Leonard Hayflick (2007), whose work will be discussed in some detail in later chapters, have claimed triumphantly that the problem has been resolved, so that ageing should no longer be regarded, as Medawar once did, as 'an unsolved problem of biology'. At the same time, discussions about the possibility of extending human longevity well beyond the present limits have captured the attention of the European and North American public, its political representatives and policymakers, motivating an increasing number of new coordinating and funding initiatives, ranging from increased budgetary allocations for the National Institute of Aging to the establishment of the European Union European Research Area in Ageing programme. It is no surprise, therefore, that books such as Thomas Kirkwood's *Time of Our Lives* (1999) have greatly captured the public's imagination.

The importance attached to ageing, as well as the uncertainty about the boundaries between old and new configurations of the phenomenon, seems inseparable from the evolution of modern forms of governance. In *The History of Sexuality* (1976), Michel Foucault argued that investment in maximizing the productive powers of the embodied, human subject was the defining feature of the modern governmental formation. He observed first the historical inversion of the importance attached to the powers over life and death, so that 'the ancient right to take life or let live was replaced by a power to foster life or disallow it to the point of death'.[2] He then argued that the latter power was mobilized in two very different, but related, modes:

> Starting in the seventeenth century, [the] power over life evolved in two basic forms; these forms were not antithetical however; they constituted rather two poles of development linked together by a whole intermediary cluster of relations. One of these poles – the first to be formed, it seems – centered on the body as a machine; its disciplining, the optimization of its capabilities, the extortion of its forces, the parallel increase of its usefulness and its docility, its integration into systems of efficient and economic controls, all this was ensured by the procedures of power that characterized the disciplines: an anatomo-politics of the human body. The second, formed somewhat later, focused on the species body, the body imbued with the mechanisms of life and serving as the basis of the biological processes: propagation, births and mortality, the level of health, life expectancy and longevity, with all the conditions that can cause these to vary. Their supervision was effected through an entire series of interventions and regulatory controls: a bio-politics of the population. The disciplines of the body and the regulations of the population constituted the two poles around which the organization of power over life was deployed.[3]

In other words, the modern governmental formation was characterized by two different forms of organizing and structuring the modes of being in relation to others. The first of these involved the articulation of juridico-political power and the material constitution of the human body, as it was disclosed by modern disciplinary discourses such as those of anatomy and physiology. The second involved instead the articulation of juridico-political power and the reproductive potential of the human species, which was operated by the administrative institutions of the modern state. Importantly, while very different in their mode of institutional organization, the relationship between the two forms of power over embodied life was a co-constitutive one, and this co-constitutive relationship was best exemplified by the discourse of 'sexuality'. As Foucault put it most pithily in the lectures at the Collège de France, which he was delivering at the time of writing *The History of Sexuality*:

> Sexuality, being an eminently corporeal mode of behaviour, is a matter for individualizing disciplinary controls that take the form of permanent surveillance ... But because it also has procreative effects, sexuality is also inscribed, takes effect, in broad biological processes that concern not the bodies of individuals but the element, the multiple unity of the population. Sexuality exists at the point where the body and population meet. And so it is a matter for discipline, but also a matter for regularization.[4]

The modern biomedical disciplines have taken shape at this point of intersection of the anatomo-political and biopolitical forms. Importantly, while the precise role of these biomedical disciplines has proved to be a matter of debate, there is little disagreement that, wherever modern governmental forms have held sway, these disciplines, when taken together, have aided the elimination of infectious diseases as major causes of mortality and then contributed to delaying the impact of chronic diseases.[5] The net effect of the conjunction has been that human longevity has increased extraordinarily during the past century. Life expectancy at birth has more than doubled since 1900, increasing from thirty to nearly seventy years, the number of centenarians has quadrupled over the last thirty years alone. Although the statistics of these demographic phenomena have also created much confusion, such as the common, but arguably mistaken, assumption that human life span has also increased, the problems presented by an ever larger fraction of the population reaching old age have coincided with a massive expansion of geriatric medicine and the medicalization of ageing generally. Moreover, while Philippe Ariès (1987), and Zygmunt Bauman (1992) after him, were perhaps mistaken when they argued that these same modern biomedical disciplines and associated governmental forms have contributed to the disappearance of death from daily life, there are good grounds for believing that these disciplines and forms have contributed greatly to the ontological evacuation of death.[6] In other words, there are good grounds for believing that death has not so much disappeared from modern daily life, as it has become instead dying, an undoubtedly productive performance around a simultaneously emptied centre, evacuated of any intrinsic meaning. At the same time, these sweeping social and cultural changes are also said to have come at a great cost. While one insurance scheme or another once defrayed the expense of defeating infectious diseases and postponing the effects of chronic diseases, it is now said that this has become difficult to sustain. It seems that the relationship between the number of socially functional individuals, usually defined as those gainfully employed and paying into these insurance schemes, and the number of biologically functional individuals has so diverged that the insurance schemes are increasingly unsustainable and require new answers. The Alliance for Aging Research thus assesses the present situation in the United States in the following terms:

> The United States is poised on the brink of a 'longevity revolution'. People are living longer, healthier lives thanks to public health advances and medical research breakthroughs. The graying of the huge baby boom generation during the coming decades will amplify this triumphant achievement, producing a society where more than one in five Americans is over the age of 65 and potentially more than one million are centenarians. Though we have seen a dramatic increase in longevity and healthy aging, our healthcare system may be overwhelmed by this explosion in numbers of older adults.
>
> By turning our attention to understanding aging and age-related diseases we can extend the healthy, active lives of older adults and limit the economic and personal burdens of an aging nation. Scientists are increasingly looking at the underlying mechanisms of aging – how we age and why we die. We are gaining a better understanding of the science of aging and are learning how to delay the onset of age-related diseases, reducing years of costly dependence on medical and long-term

care facilities and decreasing the load on an already strained healthcare system.[7]

While the Alliance for Aging Research thus translates the difficulties arising into a call on the public purse for greater investment in biomedical research, the more general response has been a massive investment in reshaping the organization of old age, aiming to better synchronize the social and biological functionality of the individual.[8]

The potential of the attempt to synchronize the social and biological functionality of the individual would seem to be limited, however, by the plasticity of ageing and death as biological phenomena. Perhaps unsurprisingly, therefore, whether the so-called 'age-specific mortality curve', which describes the surviving fraction of a population as a function of time, can be effectively squared at the individual, physiological level, so that the pathological burdens of old age are not just statistically compressed into a few months of 'catastrophic illness' just before death, but become the normal experience and common understanding of embodied life and its progress from birth to death, is a question that has greatly exercised the biomedical disciplines ever since the 1970s. In 2003, the increasing urgency of this question prompted the Milbank Memorial Fund, an influential charitable foundation that is dedicated to informing American public policy with regard to the organization of healthcare, to republish James Fries' much-cited essay on 'the compression of morbidity' twenty-two years after its first publication.[9] Whether the vertical component of the age-specific mortality curve can be eliminated altogether, which is what would obtain if every member of the population were immortal, is perhaps another matter altogether. Until relatively recently, questions posed by ageing and the possibility of immortality as strictly biological phenomena have certainly belonged to the margins of orthodox biomedical discourse, to the domain of quacks and their promises of eternal youth. There can be little doubt, however, that discussions of mortality and immortality are today attracting increasing attention. These often pit biomedical scientists invested in securing public understanding of the truth of the matter, namely that people age, that everyone must eventually die and that therefore all attention should be focused on healthy ageing and consequent good death, against the hopes invested in the promises of so-called 'anti-ageing medicine', which holds out the prospect of postponing ageing and death well beyond the present chronological limits. In the meantime, as a number of social scientists observe, the isolation and abandonment that many elderly people experience drives ever more insistent calls to reintegrate social and biological functionality by allowing these same elderly people greater control over the timing and manner of their death, by decriminalising 'assisted dying'. Some of these social scientists maintain that all these development are far from unrelated.[10]

Ageing, in sum, matters ever more to the business of governance, but it is not clear how best to address the task.

The trouble with ageing

As observed above, over the past two decades, public interest in biomedical perspective on ageing has grown considerably, but there is also equally considerable uncertainty among biomedical researchers themselves about how best to approach the subject, so testing the limits of the biomedical enterprise.

In 2005, on the occasion of the decennial White House Conference on Aging, an influential group of biomedical researchers drew attention to the minute proportion of the annual budgetary allocation for the National Institutes of Health that is dedicated to understanding the biology of ageing, calling on American legislators to reconsider their policies because 'the aging research field [is] on the threshold of a new way of thinking – shifting from a focus on specific age related illnesses to a search for an understanding of aging itself'.[11] In the United Kingdom, the Science and Technology Committee of the House of Lords has provided a forum for similar arguments. In a comprehensive report on the organization of research on ageing, the Committee has observed, on the one hand, that:

> This is ... an enormously exciting time for fundamental biological research into the causes of ageing, and into what can be done to slow the adverse effects of the ageing process and improve the quality of life of older people.[12]

On the other hand, the Committee concludes that 'most of the research on ageing and health ... is focused on specific diseases and medical conditions for which age is the single largest risk factor' and then bemoans the paucity of support for much more promising programmes of research on the 'basic processes of ageing'.[13]

These calls for greater investment in research on the 'basic processes of ageing' are fuelled in part by flamboyant biomedical scientists such as Aubrey de Grey, who claims that 'the first person to live to a thousand might be sixty [years old] already' (Grey, 2005a). Taking advantage of new modes of communication such as TED lectures and the opportunities which these offer to advance new and radical ideas both very rapidly and outside conventional mechanisms of professional scrutiny, de Grey argues that 'there is no difference between saving lives and extending lives, because in both cases we are giving people the chance of more life'.[14] Rehearsing a well-established understanding of the relationship between biology and medicine, especially as it seems to have developed in the United States, de Grey argues more specifically that the biomedical disciplines should reorient their research toward the achievement of human immortality because, thanks to developments emerging from the biomolecular sciences, particularly from within the field of regenerative medicine, human immortality is now within reach. As attests the membership of the SENS Research Foundation, the acronym SENS standing for 'Strategies for Engineered Negligible Senescence', de Grey is not alone. Echoing its founder's views, the statement of principles guiding the SENS Research Foundation sets the challenge in the following terms:

> Two thirds of all deaths worldwide, and about 90% of all deaths in the developed world, are from causes that only rarely kill young adults. These causes include Alzheimer's, cardiovascular disease, Type II diabetes and most cancers. They are age-related because they are expressions of the later stages of aging, occurring when the molecular and cellular damage that has accumulated in the body throughout life exceeds the level that metabolism can tolerate. Moreover, before it kills them, aging imposes on most elderly people a long period of debilitation and disease. For these reasons, aging is unarguably the most prevalent medically relevant phenomenon in the modern world and the primary ultimate target of biomedical research.

> Regenerative medicine can be defined as the restoration of an individual's molecular, cellular and/or tissue structure to broadly the state it was in before it experienced damage or degeneration. Aging is a degenerative process, so in theory it can be treated by regenerative medicine, thereby postponing the entire spectrum of age-related frailty and disease. But in practice, could regenerative medicine substantially postpone aging any time soon? If so, it will do so via the combined application of many distinct regenerative therapies, since aging affects the body in so many ways. Recent biotechnological progress indicates that many aspects of aging may indeed be effectively treatable by regenerative medicine in the foreseeable future. We cannot yet know whether all aspects will be, but extensive scrutiny has failed to identify any definite exceptions. Therefore, at this point there is a significant chance that such therapies would postpone age-related decline by several years, if not more, which constitutes a clear case for allocating significant resources to the attempt to develop those therapies.
>
> Unfortunately, the regenerative medicine approach to combating aging is not yet being adequately pursued by major funding bodies: only a small number of laboratories worldwide are funded (either publicly or privately) to develop therapies that could rejuvenate aged but otherwise undamaged tissues. SRF has risen to the challenge of filling this void in the biomedical research-funding arena.[15]

Attending to similar pronouncements about the powers of contemporary biomedical science, Melinda Cooper has discussed how developments within the much heralded field of stem-cell technology, developments which seem to offer the prospect of endlessly regenerating the otherwise inevitably declining capacities of the human body, fuel the promissory economy of regenerative medicine and call all understanding of the ageing body into question. Regenerative medicine, Cooper argues, is presented to the public as the biomedical answer to the limits imposed by the inevitable decay of physiological function and accompanying finitude of human existence, so transposing neo-liberal responses to the notion of limits to economic growth into the domain of biomedical enterprise. In this discursive context, the biological potential of stem cells to endlessly regenerate bodily tissues is transformed into a resource for the development of new speculative markets, backed by massive investments in biomedical start-up companies across the globe.[16] The Board of Directors and the Research Advisory Board of the SENS Research Foundation, which includes Michael West, who has contributed greatly to the understanding of the molecular control of cellular immortality and has converted such intellectual capital into its financial correlative by founding the Geron Corporation, greatly support this understanding.[17]

At the same time, however, when viewed from the perspective upon ageing that the like of the SENS Research Foundation seeks to advance, the physiological differences between the young and old are seen as a problem to be overcome. Worryingly, this understanding fuels increasingly negative attitudes toward the aged. As reports a survey of public attitudes toward biomedical research on ageing, which the Mass Observation Research Institute (MORI) conducted on behalf of the Medical Research Council and the Biotechnology and Biological Sciences Research Council:

> People did not tend to think about ageing as a life process but rather associated it with age-related problems, in particular ill health and general 'wear and tear' or 'aches and pains' as a person gets older. They tended to have a negative view of ageing and of the pressures of an ageing population on society and the environment ('There is nothing good about it'). Despite agreement that ageing happens from when you are born, people (especially the younger ones) quickly slipped back into talking about issues that face the elderly. Perhaps this points to a feeling that 'it happens to other (old) people, not me'.[18]

Such views are at odds with the political recognition that the aged and allied activists have secured since the 1980s, by insisting, sometimes quite contradictorily, that ageing is not a pathological phenomenon to be defeated. As a result of such recognition, the aged are today a group within the civic polity that is as entitled to equal protection under the law as any other. Furthermore, the understanding of ageing as a disease to be defeated seems particularly worrisome within the context of organized social welfare because it undermines the bonds across the generations that the welfare state has sought to secure. Biomedical researchers are not unaware of the tensions between this inclusive understanding of the aged and the biomedical constructions of ageing emerging from the like of the SENS Research Foundation. As the group of biomedical researchers responsible for the report issued by Science and Technology Committee of the House of Lords put it:

> What concerns us is the pervasive but often unrecognised ageist attitude of the public and the media towards diseases prevalent in old age, and the ageist approach of industry to older people as consumers. We believe the Government could do more to help combat these attitudes, directly through government departments and the NHS [National Health Service], and indirectly by their influence on schools, industry and the media.[19]

This view that ageing should not be regarded as a pathological phenomenon is absolutely central to Thomas Kirkwood's very unfavourable review of anti-ageing medicine and its promises, on the pages of what perhaps is the most renowned and respected scientific journal anywhere.[20] Yet, for all the objections to de Grey's claims that the biomolecular sciences will soon deliver something approaching human immortality, the more conservative programme for the future development of biogerontology, which the like of Kirkwood seek to advance, very powerfully helps to constitute a new, but not wholly incompatible understanding of human embodiment. If the effective contamination of youth by old age, which biogerontology advances by locating the physiological precursors of ageing and death in ever earlier phases of development, might seem to reaffirm and materially bolster the old adage that 'as soon as man is born, he begins to die', by also proposing that these are not fundamental biological processes, biogerontology also begins to lend material foundations to the growing understanding that death is an unnecessary and preventable 'life event'. From this perspective, the sometime very intense arguments around anti-ageing medicine are perhaps best described as the equally proverbial 'storm in a teacup'.

Such minor storms matter, however. Public consultations suggest that the lay public then is uncertain about the aims of gerontological research, whether it aims to extend the great gains in longevity achieved during the past century and a half or it aims instead to transform the human condition. Thus, the same survey of attitudes among the British public toward biomedical research on ageing that was quoted above also reports that:

> Scientific research into ageing is seen as important, and public funding of this area is popular, but only if this research is put into practice (and has a public benefit). Research into ageing is also seen as potentially benefiting everyone/society as a whole, based on the assumption that 'everyone is going to get old'. People tended to place a much higher value on medical research to treat the physical and mental problems associated with ageing – for example research on arthritis and Alzheimer's disease – than on research to provide cosmetic treatments. There was a feeling that scientific research has contributed greatly towards improving healthcare and is valued as a good thing ('Research is essential'). There was a recognised need for research ('We need research') and an expectation that research would lead to medical breakthroughs ('We expect a cure for almost everything'), particularly with the growing number of elderly people. There was disagreement about whether medical research was advancing quickly enough. Whilst some felt impatient at the speed of progress ('People are still dying of cancer'), others expressed concern with the speed of medical advances. There was a feeling that scientists are competing with each other in the rush to develop their ideas ('A race is going on'), whilst others described progress as like 'science fiction'. The speed of scientific advances was seen as 'frightening'. This reflected a view of scientists as not thinking about whether they should be doing things ('Scientists need to think more'). However it was mentioned that scientists are helping us to have 'better knowledge about what to do, and what not to do' in terms of healthy living. Scientific research was seen to pose important moral and ethical questions, and religious concern about human intervention in some situations. One person with a Muslim and Punjabi background expressed concern about human intervention in life and death, and expressed his belief in 'kismet' which means that 'When God says you have got to go, you have got to go'.[21]

Uncertainty about new developments emerging from studies of the biology of ageing, in other words, is substantial. The situation in the United States is not very different. The Alliance for Aging Research has reported that:

> When the public hears about aging research, they tend to think that the goal is to extend lifespan, rather than health-span. Many thought leaders said that the public most needs to understand that aging research is not aimed at extending lifespan; rather it is aimed at extending health-span. There is a common belief among researchers that the public thinks that aging research is just about 'getting people to live longer' rather than about getting people 'to live healthier longer', and it is based largely on popular media coverage of aging research.[22]

There is, in sum, a worrying divergence between public understanding, on the one hand, and, on the other hand, biomedical researchers and policymakers' own understanding of contemporary programmes for the development of gerontology and their goals.

As a result of these far-reaching uncertainties, a number of the same biomedical scientists and policymakers involved in these diverse inquiries have forged a transatlantic alliance and drafted a collective proposal to develop 'a new model of health promotion and disease prevention for the 21st century'.[23] The proposal's first author was Robert Butler, who not only coined the notion of an ongoing 'longevity revolution', but also the term 'ageism', the latter being Butler's label for the social discrimination and exclusion which the elderly have often endured and which he regarded as fully comparable to that suffered by African Americans.[24] The listing of the other authors beside Kirkwood, namely Richard Miller, Daniel Perry, Bruce Carnes, Franklin Williams, Christine Cassel, Jacob Brody, Marie Bernard, Linda Partridge, George Martin and Jay Olshansky, brings together the leading lights of British and American investment in understanding the biology of ageing and death. Unlike either the researchers associated with the SENS Research Foundation or the latter's critics, Butler and his co-signatories now call for a trajectory of biomedical development that contrasts very sharply with the many successful relationships between the laboratory and the clinic that could be said to have characterized the contemporary biomedical enterprise.

Biogerontology and biomedicine

One of the principal, if not defining, features of the biomedical enterprise since the 1930s has been the increased focusing of research on the identification of the discrete, sub-cellular components of pathways leading to pathological states. Such focusing was accentuated powerfully in the course of programmes such as the American 'war on cancer' during the late 1970s and early 1980s.[25] As Peter Keating and Alberto Cambrosio have observed in *Biomedical Platforms* (2003), where they re-examine Georges Canguilhem's argument about the relationship between biology and medicine in the light of contemporary developments, there is good reason to treat the notion guardedly, but the programme to integrate biology and medicine into a single 'biomedical' enterprise has rested ideologically on the relationship between normal and pathological states that the like of Claude Bernard helped to construct during the second half of the nineteenth century.[26] Bernard rejected categorically all qualitative distinction between the physiological states that most interested clinicians and those states that exercised zoologists and botanists, positing instead that the two were related to one another, as quantitatively different states. He thus wrote that:

> Common sense shows that if we are thoroughly acquainted with a physiological phenomenon, we should be in a position to account for all disturbances to which it is susceptible in the pathological state: Physiology and pathology are intermingled and are essentially one and the same thing.[27]

If the politics of Alzheimer's Disease, which, in many ways, has today become synonymous with old age, are at all instructive, contemporary investigations of the biology of ageing have followed this construction of the relationship between normal and pathological states whereby the two are to be regarded as 'essentially one and the same thing'. As Tiago Moreira (2009) and Margaret Lock (2013) have observed, a number of molecules associated with pathological states that are thought to be related to the chronological age of the organism have been characterized during the course of investigations in the laboratory and the clinic. Furthermore, as a result of these endeavours, a number of molecular inhibitors and modifiers are currently being tested in animal and human trials, aiming to assess their therapeutic value. Finally, clinicians appear increasingly to be willing to provide these and other life-extending treatments to older individuals. Eliding important differences, ageing, like any other disease, can now be treated, and biomedical research will thus deliver healthier and ever longer lives. Not surprisingly, the situation has captured the attention of social scientists and, following Carroll Estes and Elizabeth Binney (1989), who first coined the phrase, they now refer to the 'biomedicalization of aging' as a label for the ever wider range of social processes involved in the framing of ageing as a fundamentally biomedical phenomenon. Sharon Kaufman, Janet Shim and Ann Russ's powerful closing remarks on the situation in the United States are worth quoting at length. Effectively contributing to the intense debates precipitated, in 2003, by the publication of the Council on Bioethics' report to the President of the United States on the future development of medicine and the prospects of genetic enhancement, they write:

> The boundaries between medicine's focus on cure and its concern with life enhancement are increasingly blurred at every age. The practices of the biomedical sciences continue to move beyond the confines of disease entities and whole individuals to investigate life itself rather than disease. In doing so, they are ushering in a new genre of medicine, called by some regenerative medicine, which is part cure, part prevention, and part experimental science. Awareness of and desire for the malleability of the body and self well into late life underlies and characterizes this new kind of medicine. A major effect of biomedicalization today is that the aged body tends to be viewed now as simultaneously a diseased entity, a site for restoration, and a space for improvement. At the same time that the relevance of age for individual medical decision making is muted and denied by clinicians, the urgency about age – especially the desire to thwart its advancement by means of medical interventions – permeates the social environment. From cosmetic surgery to knee and hip replacement, from hormone replacement therapy to the newest drugs for impotence, memory enhancement, and osteoporosis, the biotechnological practices – and promises – of slowing the aging process are ubiquitous. The blurred boundaries between

management, enhancement, and staving off the effects of age lead toward the acceptance of life-extending medical procedures, regardless of age. New ways of thinking about the aged body and current and potential interventions accompanying those ideas have implications for personal, societal, and clinical responsibility that are only beginning to be explored. In its 2003 report, 'Beyond Therapy: Biotechnology and the Pursuit of Happiness', the President's Council on Bioethics warns against the pursuit of 'ageless' bodies and further temptations of life extension because those desires deflect us from realizing other human aspirations, aspirations that have nothing to do with becoming 'better than well'. Clinical medicine in consumer society contributes to 'the stretching of middle age into later life', yet, at the same time, it makes the risks of old age insistently relevant well into midlife. The well-analyzed 'age-irrelevant' society that characterized late-20th-century American social life and distinguished it from earlier historical periods has been joined, at the end of the 20th and early 21st centuries, by an additional cultural phenomenon: biomedicalized life. That is, we now have the opportunity and the (impossible) burden of 'growing older without aging'. 'The extension of medical jurisdiction over health itself' and, we would add, over life itself, renders medical intervention natural and normal, especially in late life.[28]

Having thus outlined the present historical conjunction, Kaufman, Shim and Russ then turn their attention to what critical perspective might be available that exceeds that offered by the 'biomedicalization of aging', and to the costs of not engaging in such a search for alternative ways of engaging with the ageing body. Constructing an alternative understanding of this ageing body, they write:

> It is through an ethics of normalcy that expectations about long lives and expectations about routine medical treatment come together. It is unacceptable to die at 71, or 81, or 91 if one can utilize routine medical care to stave off death and restore health. Things can be done, and the family is involved. Love is actualized often through the commitment to a longer life and by doing things to prolong life. The ancient question – what are our obligations across generations? – has not disappeared. As ever, we must demonstrate the ways in which we

care for the oldest members of our families and our society. We are demanding and being asked to share in those lives in unprecedented ways through the range of medical procedures now available. Biomedical technique provides the most powerful logic, the most pervasive method, to show our care. Demonstrating care and love for the oldest generation outside the frame of medical treatment, and then outside the rhetoric of rights and entitlement, is nearly impossible in American society. An alternative rhetoric, of non-abandonment, for example, is rarely articulated. If, as [Nikolas] Rose and others suggest, the body has become one of the most important sites for ethical judgments about life, what other framings are possible when an old body can 'benefit', unequivocally, from medical intervention? Are there ways, beyond the body and its medical treatment, to demonstrate worth and love?

A price is paid for hope and expectation invested in biomedical technique. We are only beginning to discover this price in terms of life extension in late life. As yet, the long-term ramifications, for our relationships with one another – of extending life and postponing death – are entirely unknown. In this regard we are all the subjects of a medico-ethical experiment taking place on a broadening social scale. As more and more individuals and families become involved in the stakes here, simply because we face the prospect of longer lives, one thing is certain: What we are willing to accept, in terms of interventions and obligations to those in late life, will change. We invite the gerontological community to begin to explore these issues with us.[29]

In many ways, Aubrey de Grey and the SENS Research Foundation exemplify the seemingly irresistible understanding of biogerontology and its future that Kaufman, Shim and Russ seek to contest. As noted earlier, the SENS Research Foundation observes that:

Two thirds of all deaths worldwide, and about 90% of all deaths in the developed world, are from causes that only rarely kill young adults. These causes include Alzheimer's, cardiovascular disease, Type II diabetes and most cancers. They are age-related because they are expressions of the later stages of aging, occurring when the molecular and cellular damage that has accumulated in the body throughout life exceeds the level that metabolism can tolerate. Moreover, before it kills them, aging imposes on most elderly people a long period of debilitation and disease.

The same SENS Research Foundation concludes therefore that:

> For these reasons, aging is unarguably the most prevalent medically relevant phenomenon in the modern world and the primary ultimate target of biomedical research.

Ageing, in other words, is an object of biomedical intervention such that, once each of the diseases of the ageing body is conquered, there is every reason to expect to live well beyond the current limits. This is all for the good, because there is nothing valuable in ageing and passively awaiting the actualization of one's fate. Yet, by evoking an alternative notion of 'care and love for the oldest generation outside the frame of medical treatment, and then outside the rhetoric of rights and entitlement', the advocates of such caring comportment not only overlook the agency of the older generation for which they so care, but they also bolster the notion that there exists an objectively identifiable physiological process that lies beyond the reach of all human intervention, medical or political. This in turn, only serves to animate and reinforce the power of anti-ageing medicine.

The perspective which Robert Butler and the other signatories of the call for 'a new model of health promotion and disease prevention for the 21st century' is very different to that advanced by either the SENS Research Foundation or the critics of the 'biomedicalization of aging'. Their central argument is that the progress of contemporary gerontological research is hindered by dependence on clinical definitions of the diseases most commonly associated with chronologically older individuals. Such dependence is troubling because it is an effect of a construction of the body that was forged in the nineteenth century and that no longer provides a useful way to understand the course of life from birth to death. As signatories to this call put it:

> The traditional medical approach to ameliorating modern chronic diseases has been to tackle them individually, as if they were independent of one another. This approach flows naturally

from our experience with acute diseases, where patients seek medical care for one condition at a time. In fact, applying this same strategy to infectious diseases in the 20th century helped to deliver the first longevity revolution. Although some infectious diseases have chronic effects on health (such as malaria and HIV infection), and others remain difficult to treat (including tuberculosis and most viral diseases), public health efforts to combat these diseases have made it possible for people in today's developed nations to live long enough to experience one or more of the degenerative and neoplastic diseases that are now the dominant causes of morbidity and death

Medical research worldwide has already accomplished much, and is certain to achieve more in decades to come, but its effectiveness will become limited unless there is an increased shift to understanding how ageing affects health and vitality. Most medical research teams are oriented towards the analysis, prevention, or cure of single diseases, despite the fact that nearly all of the diseases and disorders experienced by middle aged and older people still show a near exponential increase in the final third of the life span. Now that comorbidity has become the rule rather than the exception, even if a 'cure' was found for any of the major fatal diseases, it would have only a marginal effect on life expectancy and the overall length of healthy life.[30]

In other words, the clinical worldview was well suited to illnesses characterized by discrete and specific aetiologies, but it is inadequate to address the complexity of those pathological states that often characterize life in the late twentieth and early twenty-first centuries. Basically, the protracted temporal unfolding of these pathological states is so nearly coterminous with ageing that it unsettles the contrast between normality and pathology underpinning the clinical perspective. Furthermore, the contrast assumes that the two states, normal and pathological, can be juxtaposed directly, but this obscures understanding of the diverse and complex processes involved in the declining functional capacities of the organism such that the perspectives of the laboratory and clinic must be integrated with programmes of health screening and maintenance. As the signatories again put it:

> Attempts to develop preventive measures against individual

conditions related to ageing have been, for the most part, frustrating and unsuccessful. But in striking contrast, all of these conditions, and more, can be ameliorated or postponed simultaneously by well validated interventions that slow ageing. The interventions that have worked in laboratory animals are not now appropriate for disease prevention in humans. However, we believe that exploration of the mechanisms by which ageing can be postponed in laboratory models will yield new models of preventive medicine and health maintenance for people throughout life, and the same research will also inform a deeper understanding of how established interventions, such as exercise and healthy nutrition, contribute to lifelong wellbeing.

The pursuit of extended healthy life through slowing ageing has the potential to yield dramatic simultaneous gains against many if not all of the diseases and disorders expressed in later life. The most efficient approach to combating disease and disability is to pursue the means to modify the key risk factor that underlies them all – ageing itself. Pursuing an aggressive research strategy to devise interventions against ageing suitable for humans requires that it is a goal worth pursuing (it is), and that we have good leads to follow (we do), but it does not require that we know, in advance, which of the current ideas about mechanisms affecting the rate of ageing are most likely to produce effective interventions. A fresh emphasis on ageing should vastly accelerate the health, economic, and social benefits of the extension of healthy life, which we refer to collectively as the longevity dividend.[31]

In other words, by relying on epistemological structures and methodological prescriptions that are wholly incommensurable with the phenomenon studied, the existing approach cannot but fail to deliver effective prevention of those pathological states currently associated with advancing years. As such, the 'longevity dividend' resulting from the extension of healthy and productive lives into advanced age is lost. What is required is a wholly new understanding of ageing and the modes of intervention that best correspond to such understanding. Rather than pursuing the disease-specific model that has been deployed within many other branches of the modern biomedical enterprise, the focus on Alzheimer's Disease being the most glaring example, increases in health and life expectancy are more likely to be achieved by promoting research that focuses on the common biological antecedents of all those diseases that would appear to characterize the lives of the elderly today and hinder their enjoyment of the 'third age', and then developing measures that can best neutralize all precipitating factors. If there is some truth to the adage that 'as soon as man is born, he begins to die', the implications of this alternative understanding are breathtaking and it is not surprising therefore that the signatories to this call should understand it as advancing 'a new model of health promotion and disease prevention for the 21st century'.

Significantly, by offering a veritable biogerontological critique of biomedicine and advancing an expansive understanding of biogerontology and its remit, the biogerontological proposal of 'a new model of health promotion and disease prevention for the 21st century' medicine may be of some importance to contemporary endeavours to update Michel Foucault's understanding of the modern juridico-political order.

New biopolitical formations

When Robert Butler and the other signatories of the call for 'a new model of health promotion and disease prevention for the 21st century' write of 'established interventions, such as exercise and healthy nutrition', these interventions are understood as supporting molecular mechanisms that contribute to the organism's health and longevity. Such connection between molecular processes and the most mundane care for the body could be regarded as instantiating not just the reorganization of healthcare around the mode of prevention and the management of the entire life course of the individual, but also those more general developments which have led Paul Rabinow and Nikolas Rose (2006) to call for a new understanding of political subjectivity.[32]

As observed earlier, Michel Foucault proposed in *The History of Sexuality* that the modern juridico-political order was characterized by the mobilization of two different, but related forms of power over embodied existence, anatomo-political and biopolitical. Arguably, however, developments in the very biomedical sciences which these new modes of mobilizing juridico-political institutions empowered, particularly the emergence of a molecular understanding of genetic processes, are blurring the boundaries between these two forms, thus calling into question the contemporary significance of Foucault's otherwise persuasive proposition. Though complexly so, a gene is an epistemic object that has been produced at the intersection of the anatomo-political and biopolitical modes of organizing embodied existence, involving as it does claims about both the relationship between generations and the production of the organism, from embryo to the fully reproducing organism.[33] As such, the gene it is both an anatomo-political and a biopolitical object, and, since these modes of organization depend on different correlations of power and knowledge, the gene cannot but raise awkward questions about the coordination of objects, knowledge and power. Exercised by these questions, Rabinow (1996a) turns to Gilles Deleuze's reflections on the Foucauldian *œuvre*. In the course of these reflections, Deleuze (1986) argued that Foucault had adumbrated the advent of a new site of thought and deliberation after the death of Man, but that he had also failed to delineate the contours of this new site because, ultimately, he did not recognize how, just as language, within modernist literature, had broken free of linguistic designation and signification, so life was breaking free from the equation with the organism and its life course from birth to death. Rabinow concurs with Deleuze's assessment, but, wishing to eschew any metaphysical gesture of the kind which Deleuze and neo-Deleuzian philosophers seek to advance, he suggests in more historically circumscribed and nominalist fashion that contemporary social organization should be understood as structured increasingly by interventions at a biological level and in a manner such that the modern attempt to reduce social relations to a matter of biology gives way to recombination as the operative principle. Like Foucault before him, Rabinow is aware of the historical antecedents of such reduction and is concerned to articulate the novelty of the contemporary situation whereby 'biosociality' is something very different to 'sociobiology', sociobiology being the neo-Darwinian reduction of human existence greatly popularized by Richard Dawkins, in *The Selfish Gene* (1976). As Rabinow puts it, referring to both the troublesome sociobiological equations of human society and the reproductive dynamics of ant colonies, and the antecedent mobilization of genetics in the course of the far more notorious endeavours to reorganize the body politic along eugenic lines: 'in the future, the new genetics will cease to be a biological metaphor for modern society and will become instead a circulation network of identity terms and restriction loci, around which and through which a truly new type of auto-production will emerge which I call "biosociality".'[34]

For his part, Rose is also interested in Deleuze's insights into the historical specificity of Foucault's analyses, but, as a sociologist rather than an anthropologist, he is more attuned to Deleuze's views on the limitations of Foucault's understanding of governmental formations and their future than to the fate of the human subject. In light of his own early work on the historical genesis of disciplinary formations during the late nineteenth century, it is not surprising that Rose should have shown great interest in Deleuze's suggestion, in his much-quoted essay on the 'societies of control' (1990), that 'Foucault's characterization of "disciplinary society" was written at the dusk of such societies, which reached their apogee at the beginning of the twentieth century'.[35] Rose's subsequent work has evolved toward the characterization of the dispersed, individuating mechanisms governing late modern societies, which he has discussed in considerable detail in *Governing the Soul* (1990), in *Powers of Freedom* (1999), and, most recently, in *The Politics of Life Itself* (2007). For all his sympathies toward Deleuze's criticism, however, Rose, like Rabinow, remains wary of any metaphysical implications of such criticism, and, again like Rabinow, remains closer to Foucault's last thoughts in *Care of the Self* (1984). As Rose puts it:

> if the techniques for the care of the self are subjectifying, it is not because experts have colluded in the globalization of political power, seeking to dominate and subjugate the autonomy of the self through the bureaucratic management of life itself. Rather, it is that modern selves have become attached to the project of freedom, have come to live in terms of identity, and to search for the means to enhance that autonomy through the application of expertise.[36]

From the collaboration between Rose and Rabinow emerges the notion that contemporary subjects are constituted in the work of caring for their bodies, bodies that are understood as plastic assemblages of molecular components, and in their related exercise of biological citizenship and active participation in the evolution of a politics of life itself. Sharon Kaufman, Janet Shim and Ann Russ's citation of Rose's *The Politics of Life Itself* would seem to attest the resonance of such understanding of the relationship between forms of life and contemporary social formations. On the other hand, the definition of life on which this understanding is predicated also refers to physiological processes that are related most ambiguously to the organic forms usually associated with the expression of such processes. It seems quite difficult to understand fully the location of care and reflective deliberation when Rose writes that:

> The forms of being taken by contemporary neurochemical selfhood, the blurring of boundaries between treatment, recovery, manipulation, and enhancement, are intimately entwined with the obligations of these new forms of life. They are intrinsic to the continuous task of monitoring, managing, and modulating our capacities that is the life's work of the contemporary biological citizen.[37]

At the very same time, even though these words might evoke Deleuzian and neo-Deleuzian perspectives upon the production of the subject, Rose and Rabinow do not appear to be wholly persuaded. Thus, when weighing the relationship between his understanding of vital phenomena and classical definitions of vitalism, Rose writes:

> Some contemporary social theorists, taking their cue from Gilles Deleuze, and inspired by readings of [Henri] Bergson, are arguing for a novel kind of postmodern materialist vitalism. While I share many of these views, I take a rather different tack. My aim is not so much to call for a new philosophy of life, but rather to explore the philosophy of life that is embodied in the ways of thinking and acting espoused by the participants in the politics of life itself. [38]

In so distancing himself from Deleuze, Rose, like Rabinow and Foucault, would seem to remain attached to something like an irreducible notion of the subject and the organism.

Usefully, however, when Butler and his colleagues seek to draw connections between those molecular mechanisms that contribute to the organism's health and longevity and 'established interventions, such as exercise and healthy nutrition', they would seem to exemplify very powerfully Rose's and Rabinow's understanding of contemporary biopolitical subjectivity. Despite its raising a number of awkward questions about the relationship between this understanding of subjectivity and sociobiology, the background to the biogerontological proposal also helps to clarify the nature of the difficulty confronting both modes of understanding the complexities of contemporary embodied existence.

Biosociality, biogerontology and the mortal organism

There can be no doubt that the 'new model of health promotion and disease prevention', which biogerontologists on both sides of the Atlantic have sought to develop in response to the limitations of biomedical constructions of ageing, is predicated on a neo-Darwinian understanding of life that Richard Dawkins has done much to popularize. Life, on this understanding, is a matter of replication and the mechanisms necessary to guarantee the continuity of such replication. As noted earlier, Dawkins believes that:

> Individuals are not stable things, they are fleeting. Chromosomes too are shuffled into oblivion, like hands of cards soon after they are dealt. But the cards themselves survive the shuffling. The cards are the genes. The genes are not destroyed by crossing-over, they merely change partners and march on. Of course they march on. That is their business. They are the replicators and we are their survival machines. When we have served our purpose we are cast aside. But genes are denizens of geological time: genes are forever.[39]

From this perspective, the sole function of the multicellular organism is to enable the accurate and successful transmission of molecular, genetic information. Furthermore, what distinguishes one organism from another is where evolutionary history has fixed the balance between the resources allocated to replication and the resources allocated to ensure the successful transmission of the replicated information, including those allocated to the repair and maintenance of the machinery of transmission. Robin Holliday, a molecular biologist who, as will become evident in a later chapter, has played a formative role in the evolution of the proposal advanced by Butler and his collaborators, thus argued in *Aging* (2007):

> The germ line cells must remain in a fully juvenile state, free of error or defects, and ready to initiate the developmental programme when the egg is fertilised. The situation is quite different for somatic cells, since they will never be transmitted to the next generation. Their function is to provide the vehicle ... which facilitates the transmission of germ line cells.[40]

Furthermore, added Holliday:

> Animals can afford to relax the accuracy and maintenance of somatic cells, if this increases the resources devoted to reproduction. The set of devices, maintenance or repair mechanisms which are so essential for germ line cells become uncoupled from those in somatic cells to a smaller or greater extent in different animal species ... Apart from a few simple forms, all animals have bodies which do not survive. They age because in natural environments it is simply counterproductive to try to preserve the complex organisation of cells and structures that characterise these many species.[41]

Echoing Dawkins's language, the organism is an expendable vehicle whose sole purpose is to ensure the successful transmission of the germ line. On this understanding, the gains in human longevity beyond reproductive years that have obtained over the past century and more are regarded as having come at a biological cost, namely the increasing failure of the molecular mechanisms responsible for repair and maintenance. The increasing incidence of degenerative diseases in later years would appear to support the argument almost irrefutably.

In light of this conceptual background, the challenge confronting biogerontology and contemporary society is how best to reconcile, on the one hand, the dynamics of molecules and populations constituting the mortal organism and, on the other hand, the human subject's opposing desire for a healthier and ever longer life. This seems precisely the same challenge confronting Nikolas Rose and Paul Rabinow when they advance an understanding of the relationship between subject and life itself that is poised ambiguously between the organism as either site or effect of the processes associated with the politics of life itself. Consequently, understanding the genesis of the challenge confronting biogerontology and the answers that flow from such history may help to understand the fuller implications and possible answers to the problem confronting all those who, like Rose and Rabinow, work in the wake of Michel Foucault's evocation of the impending death of Man. This said, if the perspective which Butler and his collaborators seek to advance can be said to participate in an ongoing reorganization of healthcare around the management of the individual's entire life course, and if this is the very same transformation of biopolitical order to which Rose and Rabinow draw attention, it should also be noted how Butler and his collaborators have observed plaintively that the approach they seek to advance was first articulated more that a quarter of a century earlier, but 'there has been little progress towards making the necessary changes'.[42] Such complaints are of a piece with much that has been said by the Alliance for Aging Research and the Science and Technology Committee of the House of Lords. This suggests that there are some important obstacles to the integration of biomolecular and biodemographic perspectives. Consequently, if the proposed 'new model of health promotion and disease prevention' resonates with the answers to the problems posed by the contemporary intersection of the anatomo-political and biopolitical modes of organizing embodied existence, these difficulties also call for some closer attention to the implications of Gilles Deleuze's uncompromising understanding of life as having broken free from all equation with the organism and its life course from birth to death.

Genealogy and the catalogue of failure

To recapitulate, this chapter has sought to outline the many difficulties confronting contemporary biomedical understanding of ageing and death, as well as their conceptual implications for any understanding of the contemporary transformation of biopolitical order, aiming to prepare the ground for both a genealogy of the contemporary biogerontological moment and for the answers biogerontology has then sought to forge in response. Before proceeding further on this trajectory, however, some further clarification about historiographical method is in order.

Michel Foucault, as was noted in the introductory remarks, summarized the difference between genealogy and history by writing that the genealogist seeks to resist all temptation to recover forgotten developments and so effectively travel backward in time, to restore to memory an unbroken, causal chain from the past to the present moment. The genealogist seeks instead to disclose the contingency and freedom of the present moment by identifying 'the minute deviations – or conversely, the complete reversals – the errors, the false appraisals, and the faulty calculations that gave birth to those things that continue to exist and have value for us'. Consequently, the present genealogical enterprise will aim to draw out the multiple conceptual and institutional difficulties which inform the proposed 'new model of health promotion and disease prevention', seeking to eschew all search for historical origins. While the argument will sometime seem to verge on the biographical mode of historiographical representation, the figures conjured are to be regarded not as foundational agents, but as place-holders for different ways of thinking about ageing and dying, much in the same way in which 'Xavier Bichat' was a place-holder for Foucault's characterization of the historical transformation of medical knowledge which took place at some time toward the end of eighteenth century. Furthermore, the formative episodes which the like of Andrew Achenbaum (1995) and Hyung Wook Park (2008, 2012) have already outlined in some detail are far from unimportant and will figure again and again in the present account, but they will only figure as part of a strange story of dead ends, of failures to coordinate the diverse forms of knowledge and institutions associated with the names Alex Comfort, Leonard Hayflick and Thomas Kirkwood.[43]

In sum, the genealogy of the present biogerontological moment that will unfold over the next two chapters is best understood as a catalogue of failures to coordinate and integrate, and as articulating a problematization of ageing out of which the present biogerontological moment emerges.[44] Neither the past nor the present exist independently, but they instead constitute each other in each and every moment.

CHAPTER 3

The evolutionary biology of ageing and death

This chapter sets out to outline the contours of the first of two genealogical branches constituting the contemporary biogerontological moment, namely the evolutionary perspective which Robert Butler and like-minded biogerontologists have brought to bear on the question of ageing. As observed earlier, the genealogical approach seeks to articulate the contingency of the present moment by drawing attention those 'faulty calculations that [have given] birth to those things that continue to exist and have value for us'. Focusing particularly upon on developments in the United States and in the United Kingdom over the period between the 1950s and 1970s, the chapter will draw out the difficulties involved in producing an understanding of ageing and death that was consistent with the views emerging from the neo-Darwinian integration of genetics and evolutionary biology.

As was suggested in the previous two chapters, Richard Dawkins's *The Selfish Gene* (1976) has proven critically important to the popularization of a neo-Darwinian understanding of life. Significantly, Dawkins brings the introduction to his book to a close by writing that 'the central idea I shall make use of was foreshadowed by A. Weismann in pre-gene days at the turn of the century – his doctrine of the "continuity of the germ-plasm"'.[1] There can be very little doubt that the neo-Darwinian understanding of life is inseparable from the many interpretations and reinterpretations of August Weismann's contributions to the establishment of modern biology, particularly with regard to the relationship between life and death. Weismann is just as important to the evolution of modernist sensibilities about life, death and the passage of time, literary and philosophical. As Keith Ansell Pearson has observed in *Germinal Life* (1999), his fundamentally important and unequalled introduction to Gilles Deleuze's thought, Weismann's understanding of the relationship between generations was critically important to Thomas Hardy's and D. H. Lawrence's very different, if not conflicting, attitudes toward love and sexuality.[2] Deleuze himself thought that Weismann's biological propositions were just as important to understanding Émile Zola's *The Beast Within* (1890).[3]

As will become clear over the course of this chapter, when viewed from this so extraordinarily resonant Weismannian perspective, ageing was to be regarded not as a biologically significant phenomenon, but as an artificial aggregation of all the secondary and accidental consequences of selective forces acting upon the individual organism's life cycle up to the termination of reproduction. Yet, the parallel discussion of Michel Foucault's understanding of the relationship between knowledge and power will serve to advance the notion that, however intellectually coherent, this evolutionary explanation failed to gain any foothold because the accompanying dismissal of ageing as the product of cultural and social conventions could not be reconciled with the political importance of conventions such as the organization of the population into discrete and distinct developmental cohorts. Conventions, as many a sociologist has observed, are absolutely central to the effective administration of the modern polity.

Weismann, Weismannism and evolutionary biology

August Weismann's fame as a biologist is related to his reputation for having proposed first the division of the multicellular organism into germinal and somatic components, and for having then offered the radical separation of these two components as a solution to the vexing questions posed by botanists and zoologists working during the late nineteenth century about the relationship between the development of the organism and the evolution of species. Weismann is also reputed to have thus disposed of Lamarckian accounts of the evolution of species, which allowed for the inheritance of characteristics acquired by an organism during its lifetime, and to have thus paved the way for the importance attached today to the gene and cognate biomolecular sciences as key to delivering healthier and longer lives. It is not clear, however, that Weismann would have subscribed at all or throughout his life to this account of the relationship between germinal and somatic components. It is equally unclear whether this summary account of Weismann's contribution to the constitution of modern biology is not in fact a historiographical construct produced by geneticists from the 1920s onward, as they sought to reconcile genetics and evolutionary biology. Consequently, it seems that some distinction between Weismannism and Weismann, the historical figure, is desirable, if not required.[4] This is especially important because, as Manfred Laubichler and Hans-Jörg Rheinberger (2006) note, contemporary biology continues to be haunted by its debt to Weismann and Weismannism, a debt to which *Biopolitics and the Philosophy of Death* will return in a later chapter.[5]

One of the key moments in the establishment of the view that Weismann did indeed introduce a radical and unequivocal disjunction between germinal and somatic constituents of the multicellular organism was the publication of J. B. S. Haldane's *The Causes of Evolution* (1932). In this important book, Haldane mobilized Weismann's differentiation to formalize an explanatory approach whereby the biomathematical calculations of population geneticists, which sought to account for the frequency and distribution of different traits within any given biological population, became centrally important to the study of evolutionary process, almost to the exclusion of all other approaches.[6] Significantly, Ernst Mayr, biologist and leading historian of biology who hailed Weismann as 'one of the great biologist of all time', would eventually dismiss the biomathematical approach as a 'gross simplification' whereby 'evolutionary change was essentially presented as an input or output of genes, as in the adding of certain beans to a beanbag and the withdrawing of others'.[7] Whatever difficulties it posed for the understanding of evolutionary process, the approach was equally important to Haldane's promotion of research into the biology of ageing, which resulted eventually in the publication of Peter Medawar's *An Unsolved Problem of Biology* (1952) and Alex Comfort's *The Biology of Senescence* (1956).[8] *The Biology* of *Senescence* offered a twice republished, comprehensive review of the field and a programmatic statement for future development of research on the biology of ageing.[9] At the same time, however, Comfort was also very critical of Weismann. The difficulties involved in reconciling Weismann and the emerging neo-Darwinian understanding of ageing may also mean that differences such as those between Thomas Hardy's relentless pessimism and D. H. Lawrence's belief in the creative powers of life to exceed all constraint, as well as the tensions in Zola's *The Beast Within* between 'small heredity and grand heredity, a small historical heredity and a great epical heredity', speak to unresolved ambiguities, if not contradictions, in biological understanding of the relationship between life and death.[10]

Reading Weismann on ageing and death

Alex Comfort's short preface to *The Biology of Senescence* offers an uncompromising assessment of contemporary knowledge of the biology of ageing and death as less than impressive. He wrote:

> The denunciation of a subject and its current theoretical basis as 'unsatisfactory' is a relatively easy exercise – dealing with it satisfactorily is quite another matter. No biological treatment of senescence can hope to be satisfactory in the absence of a great deal of factual information which at present is not there. I have attempted to collect as much of this information as possible.[11]

True to these words, the bulk of *The Biology of Senescence* is dedicated to gathering the missing information, and from the most disparate sources, but, judging from the arguments advanced in the first two chapters, Comfort was most concerned about August Weismann's impact upon the conceptual organization of such 'factual information'. He started with the following words:

> Man throughout history, and every individual since childhood, has been aware that he himself, and those animals which he has kept in domestication, will undergo an adverse change with the passage of time. Their fertility, strength and activity decreases,

and their liability to die from causes which, earlier in life, they could have resisted, increases.

This process of change is senescence, and senescence enters human experience through the fact that man exhibits it himself. This close involvement with human fears and aspirations may account for the very extensive metaphysical literature of ageing. It certainly accounts for the profound concern with which humanity has tended to regard the subject. To a great extent human history and psychology must always have been determined and moulded by the awareness that the life-span of any individual is determinate, and that the expectation of life tends to decrease with increasing age … Every child since the emergence of language has probably asked 'Why did that man die?' and has been told 'He died because he was old'.

Interesting psychological and historical speculation could be made on the part which this awareness has played in human affairs. From the biologist's standpoint, its main importance has been the bias which it has injected into the study of senescence. The child who asks the question and receives the answer, is familiar with 'old clothes' and 'old toys'. He has always known that he, his pets, his cattle and his neighbours will become increasingly prone to breakdown and ultimate death the older they get. He has observed from the nursery that inanimate and mechanical systems also deteriorate with the passage of time. He appears at a later age to derive some degree of comfort from the contemplation of the supposed generality, universality and fundamental inherence of ageing – or alternatively from drawing a contrast between Divine or cosmic permanence and his own transience. However inspiring this type of thinking may be – and it features largely in the past artistic and philosophical productions of all cultures – its influence and its incorporation as second nature into the thought of biologists throughout history has seriously handicapped the attempt to understand what exactly takes place in senescence, which organisms exhibit it, and how far it is really analogous to the process of mechanical wear. One result of the involvement of senescence with philosophy and the 'things that matter' has been the prevalence of attempts to demonstrate general theories of senile change, including all metazoan and even inanimate objects, and having an edifying and a metaphysical cast. Prominent with among these have been the attempts to equate ageing with development, with the 'price' of multicellular existence, with hypothetical mechano-chemical changes in colloid systems, with the exhaustion induced by reproductive processes, and with various concepts tending to the philosophical contemplation of decline and death.[12]

Basically, Comfort regarded all study of the biology of ageing and death to date as prey to the kind of circular reasoning deployed usually to answer children's questions about the world around them. He added, furthermore, that these studies were vitiated by the authors' fears of dying and their habit of allaying such fears by discovering some deeper meaning to the child's innocent observation that 'he, his pets, his cattle and his neighbours will become increasingly prone to breakdown and ultimate death the older they get'.

Weismann, as the proponent of a theory of ageing which, 'though untenable', had nonetheless enjoyed 'considerable surviving influence', was the prime target of such uncompromising criticism.[13] Comfort's principal objection was to Weismann's statement that 'death takes place because a worn-out tissue cannot forever renew itself ... Worn-out individuals are not only valueless to the species, but they are even harmful, for they take the place of those which are sound.'[14] The fuller statement of Weismann's proposition is:

> Worn-out individuals are not only valueless to the species, but they are even harmful, for they take the place of those which are sound. Hence by the operation of natural selection, the life of our hypothetically immortal individual would be shortened by the amount which was useless to the species. It would be reduced to a length which would afford the most favourable conditions for the existence of as large a number as possible of vigorous individuals, at the same time.
>
> If by these considerations death is shown to be a beneficial occurrence, it by no means follows that it might be solely accounted for on grounds of utility. Death might also depend on causes which lie in the nature of life itself. The floating of ice upon water seems to us to be a useful arrangement, although the fact that it does float depends upon its molecular structure and not upon the fact that its doing so is of any advantage to us. In the like manner the necessity of death has been hitherto

explained as due to causes which are inherent in organic nature, and not to the fact that it may be advantageous.

I do not however believe in the validity of this explanation; I consider that death is not a primary necessity, but that it has been secondarily acquired as an adaptation. I believe that life is endowed with a fixed duration, not because it is contrary to nature to be unlimited, but because the unlimited existence of individuals would be a luxury without corresponding advantage. The above-mentioned hypothesis upon the origin and necessity of death leads me to believe that the organism did not finally cease to renew the worn-out cell material because the nature of the cells did not permit them to multiply indefinitely, but because the power of multiplying indefinitely was lost when it ceased to be of use.[15]

Weismann proposed, in other words, that ageing and dying were not to be explained by internal, physiological mechanisms, but by the dynamics of natural selection. Viewed from this perspective, he also seemed to maintain, worn-out and exhausted individuals were to be regarded as playing no useful role in the history of living forms, and the possibility of immortality was to be regarded as an equally useless 'luxury'. As such, natural selection was bound to eliminate both. Comfort's criticism of this explanation of ageing and death was that it 'both assumes what it sets out to explain, that the survival value of an individual decreases with increasing age, and denies its own premise, by suggesting that worn-out individuals threaten the existence of the young'.[16] In other words, according to Comfort, Weismann did not explain why, but instead simply assumed that increasing age led to impaired biological function, and then argued, not wholly coherently, that this debilitating process was of some benefit to the continued evolution of the species. Significantly, such opprobrium was directed equally toward Weismann's students, including Eugen Korschelt, whom Comfort singled out as culpable of perpetuating 'legendary and anecdotal' information about the longevity of different species.[17] There is no reason to believe that Comfort was aware of Martin Heidegger and his debt to Korschelt, but, as will be argued in a later chapter, the connection between Heidegger, Korschelt and Comfort's critique is important. In the meantime, Comfort advanced instead the notion that 'senescence is to be regarded not as the positively beneficial character which Weismann believed it to be, but as a potentiality lying outside the part of the life cycle which is relevant to evolution'.[18] More specifically, drawing on Haldane's and Medawar's biomathematical considerations on the genetics of populations and evolutionary process, the very understanding that Ernst Mayr would later vilify, Comfort maintained that natural selection operated most forcefully on those phases of the life cycle that were related to reproduction. Consequently, the phenotypic expression of any deleterious mutations, that is to say, their expression at the level of the organism, would be targeted more strongly when these mutations occurred during phases related to reproduction than it would when they occurred in any phases of life following reproduction. This differentiation, according to Comfort, would lead to an accumulation of deleterious genes whose phenotypic expression occurred in the later phases of life, eventually resulting in the genetic determination of the post-reproductive, gradual weakening of the organism commonly known as 'ageing'. As he put it:

> The declining evolutionary importance of the individual with age may be expressed in another way in the 'morphogenetic' senescence seen in mammals. At the point where a system of differential growth ceased to be regulated by forces which arose from natural selection, it would cease to be under effectively directional morphogenetic control, and would resemble an automatic control device which has run out of 'programme'. In any such system the equilibrium must be increasingly unstable. These two views of senescence, as accumulation of delayed lethal or sublethal genetic effects, and as a withdrawal of the evolutionary pressure towards homoeostasis with increasing age, are complementary, though probably only partial, pictures of its evolutionary significance. The concept of senescence as exhaustion of programme also restores a far greater unity to our definition of ageing, which includes a great many effects having little in common beyond their destructive effect on homoeostasis. All such effects fall within the idea of deterioration lying outside the 'terms of reference' of each species, as laid down by natural selection. The 'flying bomb' which failed to dive on its objective would ultimately 'die' either of fuel exhaustion, or through wear in its expendable engine. If its design had been

produced by evolution, and its evolutionary relevance ceased at the moment of passing its objective, or decreased as a function of the distance flown, both these events would be outside the programme laid down by the selective equilibrium, as they were outside the calculation of the designing engineers. Death in such an expendable system may result from one of many factors, and even ... from the consequence of processes which contribute to fitness during earlier life, such as systems of differential growth. We shall find a good deal of gerontology is primarily the study of a living system's behaviour after its biological programme is exhausted. The various evolutionary explanations of ageing already combine to offer us some idea of the reasons why this may be so.[19]

In sum, according to Comfort, the emergence of ageing and death was not the fruit of any selective advantage to the species, as Weismann supposedly maintained, but a secondary and accidental consequence of the selective forces shaping earlier phases of the individual organism's life cycle, up to the termination of reproduction. Reproduction is what the organism is programmed to achieve and it is expendable thereafter. As such, Comfort maintained, not only did studies of the biology of ageing and death focus on the wrong phase in the life cycle of the organism, but there was also an intimate and crucially important link between sex and death. While Comfort may have started to consider this link much earlier than the publication of *The Biology of Senescence* might suggest, it should perhaps come as no surprise then that Comfort is today most famous as the author of *The Joy of Sex* (1972), an enormously successful manual of instruction for the greater enjoyment of sexual intercourse, even into advanced age.[20]

Defining the phenomenon

While Alex Comfort's claims about the status of the gerontological field were uncompromising, his was in fact only one among the many competing theories about the nature of ageing that flourished within the biological and social sciences between the late 1930s and early 1950s. During these years, a variety of institutional actors became interested in the biology of ageing, partly as a consequence of the efforts to come to terms with, and respond to, the social strains that followed upon the heels of the global economic crises of the early 1930s. What is most evident throughout is that there was no consensus about either the nature of the phenomenon or the means that should be employed to study the phenomenon.

The divergence of definitions and approaches was particularly evident during formative conferences supported by two of the most active philanthropic supporters of research on ageing, the Josiah Macy Foundation in the United States and the Nuffield Trust in the United Kingdom. During the 1930s, the Josiah Macy Foundation was responsible for funding social surveys of the aged and commissioning a seminal conference on the biology of ageing, the Woods Hole Conference on the Problems of Aging. As Hyung Wook Park (2008) has documented, Edmund Cowdry's edition of the published proceedings of this conference was riven by disagreement between the various contributors, mostly around the parameters and standards by which to contrast normal and pathological ageing. Similarly, the Nuffield Trust was a key sponsor of research on ageing and is renowned for having sponsored the Survey Committee on the Problems of Ageing and the Care of Old People, whose statistical surveys brought the living conditions of older citizens to public attention. What is perhaps less appreciated is the extent of the Nuffield Trust's support for research into the biology of ageing, which started in 1940, with Lord Nuffield's personal interest in and support for research on the endocrinology of ageing, and expanded throughout the 1940s and 1950s into a substantial network of research centres across British universities and hospitals.[21] Here too, however, the uncertainties were deep. When the Nuffield Trust sponsored a symposium to showcase its achievements in this domain, the Institute of Biology Symposium on the Biology of Ageing, there was much disagreement between those who wanted to find standards by which to contrast normal and pathological ageing and those like Comfort who considered the endeavour pointless because ageing and the diseases usually associated with the ageing body were, in their view, one and the same thing.[22]

The questions raised during these diverse encounters were brought into sharpest relief during a further conference, which brought together the leading experts on the biology of ageing on either side of the Atlantic. The second meeting of the still novel International Gerontological Association was held in London in 1954. Soon thereafter, on the occasion of the first CIBA Foundation Colloquium on Ageing, the like of Peter Medawar, then Professor of Zoology at University College London, and Comfort, Nuffield Research Fellow in the same Department of Zoology, were joined in discussion of the subject by Cowdry, Professor of Anatomy at the University of Washington, and his own close associate, Albert Lansing, Professor of Anatomy at Emory University by the time of the proceedings' publication, as well as by Nathan Shock, director of the section of gerontology at the United States National Institutes of Health. According to the published proceedings of their meeting, following the chair's introduction of the concluding, general discussion, Comfort intervened in the following terms:

> I would like to put in a plea for [Professor Medawar's] definition of senescence as the increase in liability to die with advancing age. It may be proper to distinguish ageing from senescence, but in that case I think we can scrap ageing altogether and call it development, because gerontology is an entity which only comes into existence to describe a process human beings don't like, a deteriorative process, and I take it that it is senescence with which we are concerned here.
>
> Earlier in the meeting Dr. Lansing made a declaration of faith on the subject of the overall unity of the senescent process. I don't want to speak out of turn, but I'm somewhat sceptical of [the] underlying unity of any ageing process.[23]

Comfort's view that ageing was not a properly biological phenomenon and his consequent plea that ageing should be regarded as the product of cultural conventions did not go unchallenged:

> Lansing: But take the male rotifer: it is born, it has no alimentary trait and dies of starvation within twenty-four hours after fertilizing. Does he die of senescence? I'd rather put him in a special category, as a very degenerate character who starves to death in the twenty-four-hour period that he is busy fertilizing. [...] When I think of senescence I think of something that happens not to children or to infant rotifers, but to the organism that

has become an adult and then undergone some type of change, to wind up dead sooner or later. That's what I mean by senescence. The maturation of the embryo, the new born child, the adolescent, the changes with time prior to maturation, to me are not senescence.

Cowdry: Yours is the downswing of life, then.

Lansing: Yes, after adulthood has been reached. I can't define adulthood too well, and in some cases the changes that occur in adulthood are said to be improvements rather than losses.

Cowdry: You don't have to define it if you just call it the downswing, that implies that after a height you start to go down.

Comfort: Do you agree then that for various organisms the factors that contribute to that downswing tend to differ very radically from phylum to phylum?

Lansing: I'm not prepared to agree to that. I think we have special cases which bring about death, but not all death is due to senescence. […] The declaration of faith I made yesterday stems in part from the various types of survival curves that Dr. Comfort showed us. […] It would be quite a coincidence if all these processes all expressed themselves in the same way.

Comfort: Raymond Pearl plotted a survival curve for automobiles which was again the same shape!

Shock: I think the argument that because two different phenomena can be made to fit the same mathematical formulation they have common processes behind them is an extremely hazardous one.

Lansing: I said only that it's a possibility, I'm not prepared to say that we have as many kinds of protoplasm as we have species. I think there is a common protoplasm with basic properties of multiplication and growth, decline, irritability and so on, varying in detail, not in principle.[24]

Shock sought to bring the increasingly fractious exchange to a close in the following terms:

> I would agree that protoplasm is probably fundamentally much the same stuff, although we know that various tissues develop different functions, so that their enzyme systems must vary quite widely between different cells in the same animal. To that extent, I would agree that perhaps if you knew what it was that caused a cell to lose its ability to maintain concentration gradients, maintain its metabolic processes, you would be a long way toward understanding the ageing process. But it seems to me that the techniques that we have for investigating single cells are very meagre. Dr. Cowdry feels that if you take a cell out of its tissue it is no longer a cell. If we accept this position we are limited to unicellular organisms for study, but unfortunately most of these species simply divide and form two new cells so that 'ageing' fails to occur. Thus, we are faced with the problem of studying more complex animals or tissue, using both biochemical and physiological techniques. Since changes in the environment of the cell, produced by changing the diet of the animal, will often result in alterations in cellular enzymes, it seems to me that perhaps we are going to have to look at the problem of ageing from a number of different levels simultaneously and not try at the moment to conceptualize the entire problem in one framework. Prof. Medawar has approached the problem from a statistical evaluation of life tables; I am not prepared to accept this approach as the only way out of the difficulties. I think the examination of life table might be an index as to what you were doing to a process, but if you are going to explain ageing as a process I think ultimately you have to look at individuals, and perhaps the best way is to look at them from different points of view and at different levels of organization. I doubt if it would be possible to formulate a definition of ageing that would be acceptable to everybody and would cover all the aspects of the problem as it now stands.[25]

These diverse arguments can be regarded as involving the confrontation of three different perspectives. The first of these was that advanced by Medawar and Comfort, which was the perspective of population geneticists working with life tables and relying on evolutionary arguments to explain observed differences between populations or species. On this account, ageing or senescence was an age-specific aggregation of biological phenomena that were physiologically unrelated. Against this perspective was that of experimental biologists such as Cowdry and Lansing, who relied on particular organisms to model physiological phenomena that were assumed to obtain across different species. From this alternative perspective, ageing was to be regarded instead as a unitary phenomenon that occurred in all organisms at some point in their developmental cycle. The third perspective, represented by Shock, was, firstly, that the first two perspectives were equally valid and, secondly, that the individual should be regarded as the fundamental biological unit which could then be examined 'from different points of view'.

Aligning knowledge and power

The arguments at the CIBA Foundation Colloquium on Ageing illustrate the conjunction of modes of producing knowledge, the principal concern of philosophical reflection, and the contingencies of historical situation. Significantly, in a much-quoted interview, Michel Foucault (1977) summarized the relationship between his earliest work on the history of reason and his later work on changing forms of power by affirming that power and knowledge always are correlated and that the organization of the correlation into different 'régimes of truth' will vary historically.[26] He stated more specifically that:

> Along with all the technical inventions and discoveries of the seventeenth and eighteenth centuries, a new technology of the exercise of power also emerged which was probably even more important than the constitutional reforms and new forms of government established at the end of the eighteenth century. In the camp of the Left, one often hears people saying that power is that which abstracts, which negates the body, represses, suppresses, and so forth. I would say instead that what I find most striking about these new technologies of power introduced since the seventeenth and eighteenth centuries is their concrete

and precise character, their grasp of a multiple and differentiated reality. In feudal societies power functioned essentially through signs and levies. Signs of loyalty to the feudal lords, rituals, ceremonies and so forth, and levies in the form of taxes, pillage, hunting, war etc. In the seventeenth and eighteenth centuries a form of power comes into being that begins to exercise itself through social production and social service. It becomes a matter of obtaining productive service from individuals in their concrete lives. And in consequence, a real and effective 'incorporation' of power was necessary, in the sense that power had to be able to gain access to the bodies of individuals, to their acts, attitudes and modes of everyday behaviour. Hence the significance of methods like school discipline, which succeeded in making children's bodies the object of highly complex systems of manipulation and conditioning. But at the same time, these new techniques of power needed to grapple with the phenomena of population, in short to undertake the administration, control and direction of the accumulation of men (the economic system that promotes the accumulation of capital and the system of power that ordains the accumulation of men are, from the seventeenth century on, correlated and inseparable phenomena): hence there arise the problems of demography, public health, hygiene, housing conditions, longevity and fertility. And I believe that the political significance of the problem of sex is due to the fact that sex is located at the point of intersection of the discipline of the body and the control of the population.[27]

In other words, Foucault argues that, between the seventeenth and nineteenth centuries, the deployment of power changed from the mode of interdiction to a positive endeavour that aimed to enhance the productivity of the subjects governed. This latter mode came to operate at a myriad of different sites and by employing a variety of administrative techniques, the common aim being to enhance the integration of the human subject's life and the nascent capitalist system, to the benefit of the polity. Foucault closes the argument by reiterating the importance of sexuality to this strategic endeavour because it brings together the body of the individual members of the polity and the life of all these individuals, taken as a more or less integrated aggregate, the population. While the divergent perspectives advanced during the CIBA Foundation Colloquium on Ageing were not wholly unrelated to the methodological and epistemological questions posed by the increasing attention to 'demography, public health, hygiene, housing conditions, longevity and fertility', the importance is not just about the requirements of the historical context, but of the fit between such requirements and the answers offered by each of the parties involved.

Nathan Shock's ability to speak with sufficient authority to draw to a close the disagreement during the CIBA Foundation Colloquium on Ageing over the relative merits of evolutionary and physiological explanations, however temporarily, is inseparable from a fundamental shift in American funding of medical research, which the creation of the National Institutes of Health in 1946 signalled most visibly. This shift was tied to the private and public investments in research programmes to investigate the biology of cancer and heart disease that have come to define the modern biomedical enterprise, but ageing also fell within this orbit.[28] As Laura Hirshbein (2000) has suggested, from the 1930s onward, American physicians aligned themselves with growing public concern about the plight of the elderly in the wake of the financial crises during the same decade by promoting a view of old age as often associated with illness and infirmity.[29] This entailed the evocation of a normal ageing process and the definition of the geriatrician's expertise as the prevention and treatment of those diseases that disrupted the normal course. The nascent system of social security would then ensure the fullest experience of this normal course across the population, so that ageing might no longer be something to be feared and one might instead look forward to 'retirement' at some yet-to-be-determined age. Of course, the consequent opening of greater opportunities for employment among younger people served to consolidate the notion of an integrated, but chronologically stratified population. More importantly, as will be discussed in greater detail in the next chapter, especially in regard to the question posed by gerontology about the meaning of normality, Shock, as director of the section of gerontology at the National Institutes of Health, sought to secure the alignment of gerontology, biomedical research and the growing welfare state by providing the requisite standards. As Shock observed:

> With [the] increasing proportion of elderly individuals in the population, it is necessary to give some consideration to the

problem of their role in our society and the most economical methods for their care. Of first importance is the maintenance of health and vigor as long as possible ... It is obvious that, if these people are not to become an economic burden ... they must be permitted to contribute to the total economy of the country within the limits of their capabilities. In order to plan for their contribution to the economy of the country, it is essential to know not only the health status, but also the potential performance capacity of individuals in this older age group. We do not yet have enough studies of older people to make definite answers to these questions possible.[30]

Shock responded first by organizing case-controlled measurement of the physiological and functional capacities within the confines of the hospitals of the Baltimore Department of Public Welfare. He then moved on to the more ambitious Baltimore Longitudinal Study of Aging, which aimed to measure the evolution of individual physiological and functional capacities over time and which Shock was only able to develop after the President of the United States, Dwight Eisenhower, established a Federal Council on Aging. Reforms of health and welfare, as Shock himself put it in 1961, required the movement of the site of physiological research from the laboratory into the community such that 'the laboratory is, in effect, the community'.[31] Importantly, though contemporary with the longitudinal studies that characterized much American public health policy during these years, such as the Framingham Heart Study, the Baltimore Longitudinal Study of Aging was distinctive insofar as it focused on 'healthy individuals' alone, to the exclusion of all those who contracted any illnesses, aiming to thus disentangle 'ageing' and 'disease' and so secure the notion of 'normal ageing'.

The same distinctiveness of the Baltimore Longitudinal Study of Aging, as the next chapter will evince, also was the source of some difficulty for the further development of American gerontology. In the meantime, however, it is important to observe how Peter Medawar and Alex Comfort could not hope to contradict Shock's construction of the relationship between statistical and physiological perspectives because they simply could not mobilize any correlation of knowledge and power comparable to that from which Shock emerged, and so, to put the matter another way, they could not match Shock's growing institutional capital.

Medawar's and Comfort's views on ageing failed to resonate with the different construction of old age that took hold in the United Kingdom and that not only distanced gerontology from biology altogether, but also resulted ultimately in weakening the disciplinary autonomy which gerontology appeared to enjoy in the United States.[32]

Comparing the disciplinary affiliations of the participants in the CIBA Foundation Colloquium on Ageing, which brought together participants from both sides of the Atlantic, and the Institute of Biology Symposium on the Biology of Ageing is very instructive. No social scientists appear to have been invited to attend the CIBA Foundation Colloquium on Ageing, other than William Beveridge, architect of the British welfare state, and then only in an honorary capacity. On the other hand, and paradoxically, the Symposium on the Biology of Ageing was to focus on the biology of ageing, but was organized by, and attracted, a very diverse group of professional experts, some being either zoologists or botanists and others being either clinicians, psychologists or economists. The challenge that this group set for itself was to establish how their diverse disciplinary perspectives might be coordinated so as to construct a recognizably shared problem, grounded ultimately in the biology of ageing. As Frederick Le Gros Clark and Norman Pirie, the former an anthropologist and the latter a biochemist, put it in their introduction to the published proceedings:

> Our approach to the problems of senescence must necessarily be in some measure subjective, as indeed are the conceptions we tend to adopt about our own place in society when we arrive at a chronologically advanced age. How subjective our approach is, we can scarcely realize, so ingrained have become our traditional habits of thought. Most of us have words and gestures perfect as we live through the Shakespearean Seven Ages of Man. Yet a biological residue remains which is clearly entitled to be called organic ageing. What precisely accumulates or breaks down or fails to function or runs out or goes to seed, we do not know, and it may be that none of these metaphors is strictly applicable. We do not know whether we shall later find ourselves describing these experiences in neurological or biochemical or cytological terms; the chances are that no one of the biological sciences will alone prove adequate to our purpose.[33]

An economist's response to Le Gros Clark's own contribution to the proceedings, which was on working capacity as a measurement of ageing, was far less sanguine about the importance of biology to understanding the chronological organization of human life when he observed that 'while naturally biological factors must be taken into account, the reasons for fixation [of chronological thresholds] are not so much biological as financial and economic'.[34] Such uncertainty about the nature of ageing was just as important to the organization of medical structures in relation to older people. As Moira Martin has observed in her study of British geriatric medicine and the struggles to convert growing awareness of the problems confronting the elderly into supportive institutional formations, it was not through the laboratory, but through the survey that British physicians 'created a body of knowledge relating to the social, economic, and medical needs of the aged population in their own districts'.[35] This situation was not wholly surprising, given that the Nuffield Trust was renowned primarily as the sponsor of the Survey Committee on the Problems of Ageing and the Care of Old People. From the committee's perspective, the survey was the privileged tool to capture the experiences of those many people who moved out of the formally recognized workforce, remained financially solvent and did not fall ill, so remaining invisible.[36] As Joseph Sheldon, one of the organizers of Institute of Biology Symposium on the Biology of Ageing and author of *The Social Medicine of Old Age* (1948), one of the reports issued by the Survey Committee, put it, the Survey Committee was interested not just in 'those individuals who are ill and know they are ill, but have no intention of doing anything about it, [but also in] those who have never been ill, and probably never will be until their final illness'.[37] Furthermore, the increasingly powerful Medical Research Council provided no substantial, structured funding for research on the biology of ageing because there seemed to be nothing so biologically and clinically distinctive and remarkable about the patients' chronological age as to deserve the attention of specialist care.[38] In the absence of any strong political direction, funding by the Nuffield Trust, which has previously been characterized as not given to any directive approach to its distribution of funds, went to a wide variety of academic programmes that relied on equally diverse methods, though focusing primarily on the importance of changing organization of modern existence to the definition of normal ageing.[39] Importantly, this construction of the elderly as a part of the population of no particular medical interest, except when they fell ill, positioned gerontology outside the hospital, the main site of coordination for the development of British biomedicine during the second half of the twentieth century.[40] In the process, British physicians interested in the medical problems of the aged came to define gerontology as the field, in Lord Amulree's words, concerned with 'those elderly sick with social and economic problems'.[41] The consequent association between old age and that peculiarly British disciplinary integration of medicine and social science that went by the name of 'social medicine' only served to reinforce the uncertainty around the status of the elderly as subjects of social security, marginal, if not liminal, as they were to the institutions of both medicine and the social sciences.[42]

In sum, ageing cannot be regarded as being either a biological or a social phenomenon because power and knowledge always are correlated and the organization of such correlation will vary historically, creating different objects, depending on the circumstances obtaining.

The reality of standards and conventions

Against this background, it is noteworthy that Alex Comfort, during the first CIBA Foundation Colloquium on Ageing, observed bluntly that 'gerontology is an entity which only comes into existence to describe a process human beings don't like'.[43] Basically, when viewed from the perspective of geneticists working with life tables and relying on evolutionary arguments to explain differences between populations, the emergence of ageing and death was not the fruit of any selective advantage, but a secondary and accidental consequence of the selective forces shaping the organism's reproductive cycle. Consequently, when viewed from this perspective, ageing was best regarded, as Comfort put it on the occasion of the Institute of Biology Symposium on the Biology of Ageing, 'as a unity of effects rather than a unity of causes'.[44]

The view that ageing was less a biologically significant phenomenon than a matter of social conventions resulting in the constitution of a seemingly unified and coherent segment of the population that was particularly prone to illness and death was difficult to articulate with the institutions of the modern state. As Luc Boltanski and Laurent Thévenot (1991) observe, all forms of collective action require standards and conventions concerning the operation of such standards. Such standards and conventions matter, and so much so that to deploy them to best effect, considerable energy must be invested in establishing their objective characteristics, in making them real. As a result, one must speak of a veritable 'regulatory objectivity'.[45] In a similar vein, Judith Treas argues that there is no necessary reason to organize the institutions of social care and economic security around chronological age, but, for historically contingent reasons, age has become the pivotal variable for the segmentation, organization and administration of the population. As Treas puts it, 'when age is the basis for important standards, it is necessary to have standards for age'.[46] As such, the greater or lesser reality of ageing as a biological phenomenon is not nearly as important as is the weighing of what is to be gained from calling into question the agreement that age matters.[47] Against this background, it is important to note how, despite the progressive credentials which Comfort and the proponents of the evolutionary perspective on ageing and death enjoyed, they could also be regarded as arguing that ageing was peculiar to those species that can survive long after reproduction, thanks, for example, to the social mores that call on younger humans to help and protect the elderly, so much so that ageing might be said to be a 'disease of civilisation'.[48] As such, like the sociobiologists who followed in their steps, the proponents of this explanation could also be said to have regarded the nascent welfare state not as the answer to the problems posed by increasing numbers of aged citizens, but as part of the problem. It should come as no surprise therefore that, during in the 1960s and 1970s, when social medicine lost its always precarious institutional support within both the Medical Research Council and the National Health Service, British research on the biology of ageing, as a distinct and autonomous area of research, lost all residual disciplinary legitimacy and political support.[49] Zhores Medvedev's recollection of Comfort's laboratory, upon visiting him in 1973, is telling. He writes:

> Alex invited me to visit his laboratory at the Zoology Department of the University College. I expected to find a large research

center ... I was surprised to find that it did consist of Comfort's office only, which was at the same time the Editorial Office of *Experimental Gerontology* as well, the first journal of biogerontology which Comfort had founded in 1965.[50]

Comfort, always the anarchist, if not the libertarian, emigrated to the West Coast of the United States soon after Medvedev's visit. In other words, despite the fact that, at the time of the CIBA Foundation Colloquium on Ageing, the evolutionary perspective on ageing was a wholly British and very vibrant theoretical perspective, conceptual and institutional dynamics had worked together to progressively disconnect biological explanations of ageing from any public debates and programmes to address the 'problem of old age'.

As David Armstrong has noted, the place of ageing within the epistemic economy of the British biomedical enterprise, which Comfort sought to advance, was contradictory insofar as it was expected to derive its intellectual legitimacy from the emergent field of gerontology, which regarded ageing and death as natural processes, and its practical legitimacy from clinical medicine, which distinguished between the physiology of the socially productive adult and abnormal, pathological states, which included both childhood and ageing.[51] As such, gerontology was saddled with two contradictory notions of normality. The rise and fall of physiological capacity was to be regarded as the normal course from birth to death, but it was also important to distinguish between normal and abnormal ageing, the latter being characterized by the very processes of decline manifest during the normal course from birth to death. Viewed from this perspective, the attempt to associate the biology of ageing with the growing political concern about the 'burden of an ageing population' by establishing the field of biogerontology can be viewed as not just exemplifying, but also as shining too bright a light on the tensions and contradictions that have characterized the biomedical enterprise from its very inception.[52] This perhaps is the context against which to weigh Peter Medawar's complaint that 'those anxious about the possible malefactions of research on ageing should take comfort from the fact that the great public and private agencies are not competing with each other in their endeavours to support research on ageing'.[53]

From dead ends to founding figures

Today, Alex Comfort is perhaps more famous as the author of *The Joy of Sex* than as the author of *The Biology of Senescence*. As noted earlier, however, the two facets of Comfort's work may not be wholly disconnected.

Robert Goff, in his study of Comfort's work, makes much of Comfort's recollections of his medical studies, which juxtapose ageing and sexuality:

> I recall being uneasy as a student when teacher after teacher explained that therapy was pointless or operation unnecessary because the patient was old, had had his time, could not expect miracles, and so on. Some moved the age of expendability far forward: at fifty five or sixty the genital system became expendable ("he or she won't need it now"); at sixty five further therapy was to be limited to encouragement. The image of later life was that of the well-behaved, asexual, uncomplaining subject, patiently awaiting the next world, to be kept as a pet if cheeringly vigorous, if not, to be jollied and avoided.[54]

This said, Goff is all too aware that these very humane thoughts about the fate of the aged can also be construed as reinforcing a growing and unsettling equation whereby sexuality regained is ageing denied. Today, as Stephen Katz and Barbara Marshall observe with respect to this equation,

> Older individuals must cope with the impossible burden of growing older without aging, with a fundamental part of this burden attributed to the maintenance of sexual functionality and 'fitness' …
>
> Sexual fitness has become fashioned as a pivotal sex/age body-problematic … around which the cultural ideals of optimal lifestyles, constant activity, and successful anti-aging have coalesced and superseded the historical concerns of an earlier era around population, nation, and regeneration. Central to this problematic has been the development of the posthuman body.[55]

Such conjunction of ageing, sexuality and the evolution of the post-human body is no accident. Comfort was wont to mobilize a basic psychoanalytic approach to conjure an erotic drive that would overcome fear of decline and death, but it was an approach deaf to Sigmund Freud's *Beyond the Pleasure Principle* (1922), where the erotic and death drives are equally important to the construction of the subject. From the perspective of evolutionary biology, which Comfort sought to advance, ageing and death were to be regarded as derivative, secondary phenomena of no positive value, not even to the species, the latter being what August Weismann was said to have maintained. Thus, for all the sympathy evoked, ageing and the confrontation with death were of little importance, and the erotic drive was to be understood literally as life's strategy to free itself of the constraints imposed by the mortal body, what Comfort, in his reflections on the 'biology of religion', called the 'homuncular perspective'.[56] While Comfort's ambition to align human existence and the dynamic of life freed of all bodily encumbrance was very much part of the counterculture, to which Comfort himself contributed signally, it may also be equally important to the evolution of the post-human imaginary, not just as a matter of philosophical reflection, but also as a spur to technological innovation.[57]

Linda Partridge, founder and first director of the Institute of Healthy Ageing at University College London, as well as signatory of the transatlantic proposal of a 'new model of health promotion and disease prevention for the 21st century', which was discussed in the previous chapter, is perhaps heir to Comfort's vision when she sets the members of her laboratory to work on the identification of the genetic mechanisms involved in the determination of senescence, aiming to slow this process and to thus postpone the death of the human organism. As Partridge puts it, echoing Comfort, the task is 'to discover genes and mechanisms that determine the rate of ageing [and] the physiological mechanisms that force organisms to make trade-offs, such as that between … high reproductive rate, on one hand, and slow ageing, on the other'.[58] Significantly, the Institute of Healthy Ageing today positions itself not just as the proud heir to Peter Medawar and J. B. S. Haldane, but also to Comfort's recently rediscovered contributions to both biological understanding of ageing and sexuality:

> University College London has a long and distinguished tradition of work on the biology of ageing; indeed the roots of the modern evolutionary theory of ageing lie in the ideas of J. B. S. Haldane and Sir Peter Medawar both of whom worked at UCL.
>
> During the 1950s and 1960s, Alex Comfort (1920–2000), famously known as the author of the *Joy of Sex*, carried out his pioneering experimental research into ageing in the UCL Department of Biology. His landmark work, *Ageing: The Biology of Senescence* (1956) [sic] helped to stimulate popular interest in the subject of biogerontology.
>
> The Institute of Healthy Ageing follows in this tradition and continues to carry out ground-breaking research on the biology of ageing and ageing-related diseases.[59]

In other words, history matters, even if it sometimes has to be rewritten to displace 'faulty calculations' and so constitute uninterrupted lineages, a process that will be examined much more closely in the next chapter, after the fullest consideration of a second set of conceptual and institutional misalignments.

CHAPTER 4

Molecularizing the biology of ageing and death

As was noted at the outset of the last chapter, the 'faulty calculations that [have given] birth to those things that continue to exist and have value for us' were more than one. The second of these miscalculations concerns the molecular perspective which contemporary biogerontologists have also brought to bear on the question of ageing, especially how this perspective is to be applied to fix the fraught boundaries between normal and abnormal states.

During the 1930s, Charles Lindbergh, the famed aviator, and Alexis Carrel, another recipient of the Nobel Prize for Physiology and Medicine, this time for contributions to the development of surgery, and one of the leading biologists of the day, set out together to conquer human mortality.[1] Whatever the merits of the stories told about this extraordinary encounter, which so captivated the American public that *Time* featured the two participants in one of its many famous covers, the shared and so visible ambition rested on Carrel's widely credited claim that, under artificial conditions, biological cells could be made to reproduce forever. Death, in other words, was not to be regarded as an inescapable feature of life, and talk of achieving immortality was not to be treated as utterly nonsensical. Thirty years later, a far less renowned Leonard Hayflick challenged such understanding by arguing that the cell lines on which Carrel based his famous claims could in fact only undergo a fixed number of subdivisions. This development is nowadays considered to have been fundamentally important to the development of contemporary views on the molecular mechanisms involved in processes of ageing and dying, so much so that Hayflick earns the pioneering role in Stephen Hall's *Merchants of Immortality* (2003), a popular account of such development and the accompanying propositions that human longevity can be extended well beyond the current limits.[2] Yet, if the notion that the cell lines can only undergo a fixed number of subdivisions, up to what is now called the 'Hayflick limit', is regarded as a fundamentally important contribution to understanding the plasticity of the processes involved in ageing and dying, this notion was in fact slow to be recognized as such. Much has been said by the like of Jan Witkowski (1978, 1987) about the ways in which the idea that ageing and death were determined at the cellular level of biological organization went against the institutionally embedded understanding that these phenomena were determined at organizational levels above the cell. Focusing on wider developments in the United States during the 1960s and 1970s, this chapter will instead draw attention to the way in which the notions that the life cycle of the cell is finite and that the stages of such cycle are determined at a level below the cell emerged from the tensions between the development of standardized materials for the production of vaccines for human use and the protocols of experimental oncology. Furthermore, while the resultant understanding of the cell could be reconciled with practices in these two very important domains of the contemporary biomedical enterprise, it was at odds with the notion of normal ageing. During the 1960s, Nathan Shock, the leading American gerontologist of the day, sought to develop this notion and so secure institutional support for biomedical research on ageing within the expanding organization of the United States National Institutes of Health. Shock and others seem to have regarded Hayflick's claims about the life cycle of the cell as useful to this end since they promised to open the way for the creation of new experimental systems with which to examine the physiology of normal ageing. Hayflick himself, however, did not regard his work as endorsing any notion of normal ageing, and his work was also very much at odds with the tradition of contrasting physiological states to establish the norm as the foundation of all properly biomedical research, a principle reaffirmed in what was perhaps the key text in the formation of American gerontology, Edmund Cowdry's *Problems of Ageing* (1939). Finally, institutional support for biomedical research on ageing could only be secured with the emergence of an institutional understanding of ageing that, despite all the objections of contemporary gerontologists and many political activists, constituted ageing as a fundamentally pathological state and motivated the current, extraordinary institutional investment of resources in defeating Alzheimer's Disease.

Importantly, in the course of outlining the many miscalculations and misalignments involved in the emergence of a molecular perspective on questions of ageing and death, this chapter will also consider the significance of such developments for Peter Keating and Alberto Cambrosio's analysis of the contemporary biomedical enterprise, as it is advanced in *Biomedical Platforms* (2003). While ageing and death have become objects of biomedical inquiry, they also seem to defy any organization into the single and coherent 'space of representation' which Keating and Cambrosio evoke when revisiting Georges Canguilhem's formative analysis of the fraught relationship between biology and medicine. In other words, there seems to be something about ageing and death that is refractory to any easy alignment or realignment of the normal and pathological states that lend the title to Canguilhem's *magnum opus*, *The Normal and the Pathological* (1966).

Revisiting the normal and the pathological

Georges Canguilhem, an extraordinarily influential teacher, was greatly concerned to explore the conceptual enframing of vital phenomena, and he was particularly interested in the conceptual difficulties involved in the development of biomedical understanding of the relationship between normal and pathological states.[3] In *The Normal and the Pathological*, Canguilhem observes how, during the second half of the nineteenth century, Claude Bernard had posited that:

> Common sense shows that if we are thoroughly acquainted with a physiological phenomenon, we should be in a position to account for all disturbances to which it is susceptible in the pathological state: Physiology and pathology are intermingled and are essentially one and the same thing.[4]

Notions such as this, Canguilhem argued, were critically important to the hierarchical ordering of biology and medicine to which the biomedical enterprise is heir, but they were also fundamentally flawed. Canguilhem also wrote:

> With Bernard as with [Xavier] Bichat, [François] Broussais and [Auguste] Comte, there is a deceptive mingling of quantitative and qualitative concepts in the given definition of pathological phenomena. Sometimes the pathological state is 'the disturbance of a normal mechanism consisting in a quantitative variation, an exaggeration or attenuation of normal phenomena', sometimes the diseased state is made up of 'the exaggeration, disproportion, discordance of normal phenomena'. Who doesn't see that the term 'exaggeration' has a distinctly quantitative sense in the first definition and a rather qualitative one in the second? Did Bernard believe that he was eradicating the qualitative value of the term 'pathological' by substituting for it the terms dis-turbance, dis-proportion, dis-cordance?[5]

Canguilhem mobilizes this elision of difference between qualitative and quantitative distinctions to insist upon the fundamental irreducibility of the clinical perspective. There is, in other words, something about the living body that exceeds and escapes biological determination, a claim of some importance to two of Canguilhem's most famous students, Michel Foucault and Gilles Deleuze.

In *Biomedical Platforms*, Peter Keating and Alberto Cambrosio argue very persuasively that Canguilhem's so powerful argument may be the product of exclusive focus on concepts and discourse, which then leads to all-too-ready agreement that modern biomedical knowledge leads to the fragmentation of the body. They write:

> Some, taking their cue from historians such as Michel Foucault, have described the dispersion of 'the clinical gaze' through a fragmentation of the patient's body, whereby the latter 'is no longer localized in the discrete, integral body of the actual patient', but, rather, is distributed (figuratively speaking) among a number of different specialties, and (in a literal sense) simultaneously present as a set of samples in different sections of a hospital ... This fragmentation of the body is, moreover, held to mimic social fragmentation, as instantiated in the increasingly complex division of labor.

They then add that:

> To speak of a fragmentation of the patient is to lose sight of all the work that goes into keeping everything together, into making sure that, for instance, the sample that has left the body will rejoin it in terms of meaningful (for the task-at-hand) results.[6]

Keating and Cambrosio go on to argue for greater attention to the material practices involved in the articulation of concepts and discourses, because, if the biomedical sciences have in fact overcome the historical division between medicine and biology, it has been thanks to the production of novel material entities, which 'without eliminating the distinction between the normal and the pathological, function within common experimental systems and a single space of representation'.[7] It is not clear, however, that the cell lines which Leonard Hayflick produced and the course of their complex movement between the manufacture of standardized materials for the production of vaccines for human use, the setting of protocols of experimental oncology and the renewal of experimental gerontology, can be viewed from this perspective and its emphasis on pragmatic, if not institutional, mechanisms of integration. As Laurent Thévenot (2009) observes, in his further reflections on the notion of regulatory objectivity, which *Biomedical Platforms* does much to advance, the agreements to which Keating and Cambrosio refer are impressive local achievements, but the opening of connections between one locality and another is also critical to the process of innovation and such movement often comes at the cost of destabilization, even fragmentation.[8]

Setting a new programme for gerontology

In 1972, the Gerontological Society of America bestowed upon Leonard Hayflick the Robert Kleemeier Award, which was granted annually 'in recognition of outstanding research in the field of gerontology'. Strikingly, Hayflick opened the Kleemeier Lecture, which he delivered a year later, by positioning himself as an outsider to gerontology, confessing that he felt 'somewhat uncomfortable in accepting an award for work which at the outset was undertaken with the biology of aging farthest from my mind', which then enabled him to play the role of the dispassionate critic who was unencumbered by disciplinary 'old dogmas'.[9]

During the course of his lecture, Hayflick, the reluctant gerontologist turned historian, traced first how, between 1961 and 1965, he had challenged the widely accepted understanding that somatic cells, those cells not involved in the reproduction of the multicellular organism, were potentially immortal. He had done so by demonstrating experimentally that the number of replications that cells could undergo was in fact limited and that the limit was fixed at the nuclear level. By the late 1960s, he argued, he and others had turned this limitation into a practical resource for the development of the novel field of 'cytogerontology'.[10] Along the way, Hayflick noted sarcastically the unreasonable dogmatism of those who refused to contemplate that the universality of 'ageing and death' might extend to the cells constituting the whole organism and that such dogmatisms might be a matter of interest to the 'psychologists and psychiatrists' in the Gerontological Society of America.[11] On a more serious note, he asked what might be the implications of the distinctions between ageing and longevity that were emerging from the examination of differences between cell lines *in vitro* and *in vivo*, answering that:

> The first is that the primary causes of age changes can no longer be thought of as resulting from events occurring at the supracellular level, i.e., at cell hierarchies from the tissue level and greater. The cell is where the gerontological action lies. I believe therefore that purely descriptive studies done at the tissue, organ and whole animal level, as they pertain to the biology of aging, are less likely to yield important information on mechanism than studies done at the cell and molecular level.[12]

This veritable epistemic shift, according to Hayflick, was absolutely necessary. As he observed, genetics, molecular biology and the study of differentiation were being taken up by an ever larger number of biologists, but, despite the universality of death, the subject remained a matter of interest to 'a minuscule number of biologists'. This, he argued, was because ageing and death had become the domain of 'eccentrics and charlatans' who promised to deliver immortality and the 'sensational press and television' who lent credibility to such promises.[13] More importantly, Hayflick also maintained that if the field was open to such misappropriation, it was because too many biologists considered that these phenomena were too complex to be amenable to experimental investigation and the consequently inadequate understanding of the relationship between the two saddled clinicians with the modern medical dilemma, 'using every means for prolonging the terminal stages of disease in the name of prolonging life but at the expense of continuing the agony of certain death'.[14] Gerontology, he argued, should therefore appropriate the study of human longevity, an appropriation rendered possible by the development of cytogerontology, and not as an end in itself, but rather as a means to understand the degenerative processes that led cells *in vivo* never to reach the longevity observed *in vitro* and so enhance the possibilities of extending 'our most vigorous and productive years'.[15] Furthermore, if this ambitious programme to identify the 'fundamental biological causes of aging' echoes Nathan Shock's formative construction of gerontology, Shock's programme was underpinned by the need to distinguish between normal and pathological ageing, but, on Hayflick's vision, this would become a particularly difficult challenge:

> One is forced to conclude that if all disease related causes of death were to be resolved, then the aging processes would present some clear physical manifestations well in advance of death itself. The challenge, of course, is to separate disease-related changes from the basic biological changes that are a part of the aging process. Since fundamental aging processes most certainly contribute to or allow for the expression of pathology, then the two concepts may be so closely intertwined as to make any clear distinctions a futile exercise in semantics.[16]

In sum, by advancing the notion that the biological organism should be regarded as composed of parts, each subject to ageing and dying, Hayflick's programme for the future development of gerontological research held out the promise of aligning gerontology with the growing field of molecular biology and its own ambition to reconfigure the entire domain of biology and medicine, but at a price, namely the abandonment of any distinction between ageing and disease.

The complexities of biological standardization

It is not easy to understand how Leonard Hayflick was ever able to stand before the Gerontological Society of America, as the latter bestowed upon him one of its principal prizes and looked to him as the new guiding light, and, at the very same time, he could seek to set a new programme by openly challenging the discipline's principal and defining tenets. Part of the answer involves the process of standardization.

Angela Creager (2002) and Jean-Paul Gaudillière (2002) argue that the standardization of experimental materials was fundamentally important to the historical alignment of biological research, medicine and American society during the middle decades of the twentieth century. More specifically, Creager and Gaudillière maintain that the standardization of experimental materials such as the mouse and the virus was important to the development and expansion of the research programmes of the National Cancer Institute and the National Foundation for Infantile Paralysis to conquer cancer and poliomyelitis, in large part by mediating connections and the movement of resources between the worlds of academic research, industrial innovation and public policy. In other words, according to Creager and Gaudillière, the mouse and the virus were devices that served to coordinate disparate conceptual innovations, materials entities and social institutions.[17] As Hannah Landecker has argued in *Culturing Life* (2007), the same argument can be advanced with respect to the human cell, and this thanks in part to Hayflick's work. It is not clear, however, how the human subject came to be integrated into these operations of standardization, thus motivating the like of Ilana Löwy (1998) to study the movement of materials from bench to bedside, and Harry Marks (1997) to draw attention to the importance of epidemiological research as a form of experimental research and political intervention that aimed to secure such integration.

As the next two, parallel sections of this chapter will seek to establish, understanding Hayflick's peculiar position calls for greater attention to the convergence of those developments which, from the early 1960s onward, led Hayflick to challenge existing understanding of the human cell and their cultivation, and the evolution of the study of human communities, particularly as advanced, at the very same moment, by Nathan Shock and the Baltimore Longitudinal Study of Aging.[18]

Cells, between viruses and cancer

In many ways, Leonard Hayflick's early career exemplifies the alignment of academic institutions, industrial research and material practices which has drawn the attention of many historians of the biomedical sciences and their development in the United States during the 1950s and 1960s.

In 1956, after a brief, but formative period of employment in the laboratories of a major pharmaceutical company, Sharp & Dohme, followed by doctoral studies at the University of Pennsylvania, Hayflick undertook further, post-doctoral training in the serial cultivation of cells derived from human tissue. Importantly, he did so with one of the leading cytologists of the day, Charles Pomerat, at the very time when colleagues in the same laboratory were seeking to establish the normal human chomosomal complement.[19] Two years later, Hayflick moved to the Wistar Institute in Philadelphia, where he was charged by Hilary Koprowski, formerly of Lederle Laboratories and new director of the Wistar Institute, with the construction and maintenance of human cell lines. Famously, in the wake of the fatalities that had followed Cutter Laboratories' release of a vaccine against the poliomyelitis virus that had been produced following Jonas Salk's controversial protocol and the ensuing race to develop a safer alternative, the like of Albert Sabin focused on the treatment of the virus. Those like Koprowski focused instead on the medium of cultivation and production, especially because investigation of Salk's vaccine had revealed the presence of a previously unknown virus in the simian cell cultures used to produce the vaccines, which was eventually named SV40, further heightening fears about the safety of the vaccines. Koprowski came to regard human cell lines as potentially very important, but their use was fraught with difficulties, mostly because the varying number of chromosomes found in human cell lines under cultivation caused these cell lines to resemble very worryingly cancerous cell lines. As Hayflick endeavoured to construct human cell lines that were free of all abnormalities that might call into question the safety of their use as a medium for the production of a new vaccine, and so help to position the Wistar Institute as a leading centre for biomedical research, he focused his attention on why cell lines, after a few months of cultivation, stopped subdividing and died. While it was commonly assumed that cultured cell lines could be subdivided indefinitely, questions about such immortality had been raised throughout the 1950s, as these lines came into ever wider use, and it was becoming increasingly difficult to attribute the observed mortality to either unforeseen pathological events or to the technical difficulties involved in the maintenance of indefinite reproduction.

In 1961, Hayflick and Paul Moorehead, who had also migrated from Pomerat's laboratory to the Wistar Institute, sought to publish the results of their investigations into the observed mortality of the cell lines. Drawing on their training in Pomerat's laboratory, Hayflick and Moorehead came to believe that the number of subdivisions a cell line could undergo was in fact limited and that subdivision beyond the limit they proposed resulted in the production of increasingly abnormal cells, the evidence being the increasingly variable chromosomal complement found in the later generations. Importantly, normality here was not to be defined as the modal chromosomal number, but in accordance with the cytological procedures established by the researchers in Pomerat's laboratory. In other words, Hayflick and Moorehead were proposing a properly biological, as opposed to statistical, standard to distinguish between normal and abnormal reproduction, the latter resulting in phenomena similar to those found in cancerous tissue. Significantly, if Hayflick and Moorehead believed that the lines which they had developed disposed of contemporary objections to the use cultured human cells to produce safer vaccines against the poliomyelitis virus, their findings threatened to disrupt the coordination of virus, cell and cancer which supported the increasing importance of virological research within the expansive domain of experimental oncology. As Jean-Paul Gaudillière (1994; 1998) has observed, by the middle of the 1950s, the American federal government was beginning to invest very heavily in research aiming to understand the role of viruses in oncogenetic processes. If cell cultures were critically important to the experimental manipulation of the viruses thought to be responsible for oncogenesis, however, they were also regarded as a passive and wholly unproblematic medium such that any observed cellular change could be treated as the effect of some putative viral agent. Effectively, Hayflick and Moorehead called into question this experimental system by suggesting that the same change could be regarded equally plausibly as an effect of the cell's life cycle. It is not surprising therefore that Hayflick and Moorehead's findings were rejected by experimental oncologists, most notably by Peyton Rous, editor of the prestigious *Journal of Experimental Medicine* and leading exponent of the viral theory of oncogenesis, so much so that the report was not accepted for publication in the *Journal of Experimental Medicine*. Hayflick and Moorehead's article eventually appeared on the pages of *Experimental Cell Research*, the much more recently established official organ of the International Society for Cell Biology.[20]

If the programme of the Wistar Institute to become the leading centre for the development of the emerging alliance of biology, medicine and the pharmaceutical industry was thus blocked by the material and institutional organization of experimental oncology, in 1965, Hayflick renewed the challenge by publishing a comprehensive restatement of the earlier findings, this time casting aside any uncertainty about nature of the phenomenon as possibly being an artefact and asserting that 'the finite lifetime of diploid cell strains *in vitro* may be an expression of aging or senescence at the cellular level'.[21] He explained that 'a cellular theory of aging is generally considered unacceptable because of the apparent 'immortality' of cell cultures', citing the leading biologists of the day and then correcting their collective view by noting the many ways in which indefinite cellular subdivision was associated with abnormalities so closely associated with cancer that the transformation into an indefinitely reproducing cell line should be 'regarded as oncogenesis *in vitro*'.[22] Hayflick thus turned the proverbial table on the virologists who had originally dismissed Hayflick and Moorehead's argument, but he also offered an equally proverbial olive branch by suggesting how the acceptance of his findings and explanation might reinforce the viral theory of oncogenesis. He wrote:

> Reproducible conditions for inducing such alterations [from limited to unlimited subdivision] would be a most powerful

tool for the study of the in vitro conversion of normal human diploid cells to cancer cells. Recently, it was discovered that the infection of primary cultures or human diploid cells with the virus SV40 could provide these conditions.[23]

Furthermore, noting the coincidence of the transition from limited to unlimited subdivision and abnormal chromosomal complement, as well as the nature of the survival curve once the rate of cellular subdivision entered its declining phase, Hayflick suggested more tentatively that the key to the regulation of cellular longevity and its overturning in the course of oncogenesis might be processes and events occurring at the nucleic level. This argument brought Hayflick into alignment with the growing convergence of molecular biology and experimental oncology, which eventually resulted in Hayflick's appointment to a chair in microbiology at Stanford University.[24] More immediately, however, the argument did not alter the position of the United States Public Health Service's Division of Biologics that human cell cultures were an unsafe medium for the production of vaccines, despite Hayflick's not-so-subtle intimation that the virus found in the preferred simian cell cultures was regarded increasingly as involved in oncogenetic processes and so reversing Sabin's argument against their use. The Division of Biologics continued to maintain its position even after an outbreak of haemorragic fever in 1967, which killed a number of German workers at the pharmaceutical company Behringwerke AG who were handling monkeys destined for the production of the company's vaccines. These deaths figured prominently during an international meeting on cell culture and the production of vaccines, which was organized by the Division of Biologics and on which occasion Hayflick restated his case for the use human cell lines, attracting support among a number of researchers other than Sabin and representatives of the Division. By the late 1960s, the issue divided American pharmaceutical companies, which faced both rising costs of litigation and increasing uncertainty over the safest method of producing vaccines for human use, especially as foreign companies opted to use Hayflick's WI-38 human cell lines and sought approval to sell their vaccine on the American market. In 1971, the situation precipitated the organization of a series of hearings before the United States Senate's Subcommittee on Executive Reorganization and Government Research, which would eventually result in the Consumer Safety Act of 1972 and on which occasion Hayflick contributed very visibly to the transfer of the Division of Biologics from the National Institutes of Health to the Food and Drugs Administration. He did so by charging that by failing to license the use of human cell lines, a self-interested Division of Biologics had endangered the safety of the American public.[25]

In sum, throughout the 1960s, Hayflick's first and foremost concern seems to have been to establish that the mortality of human diploid cells lines was so normal as to make these lines the perfect medium for the testing and production of vaccines. The debates over the biology of cultured cells, to which Hayflick contributed signally, positioned him as a highly visible figure in the political debates at the intersection of academic and industrial research, which characterized the renewal of the American pharmaceutical industry from the 1970s onward.

Coordinating human lives and the life of the cell

That Leonard Hayflick's first and foremost concern was to establish the importance of human cell lines to the production of vaccines lends credence to the confession in 1973 that his work 'was undertaken with the biology of aging farthest from [his] mind', but it also raises questions about how exactly his work then came to be regarded as a contribution so important to the progress of gerontology that the Gerontological Society of America should bestow upon him the Robert Kleemeier Award. Answering this question requires some understanding of contemporaneous, independent and orthogonal developments within American gerontology, or, in other words, some understanding of coordination as a highly contingent process.

There can be little doubt that, during the 1950s and 1960s, Nathan Shock, director of the gerontology branch in the National Institutes of Health, played a pivotal role in integrating gerontological research within the changing institutional organization of American public funding for biomedical research. As Betty Lockett has observed, following John Kennedy's election to the presidency of the United States, the Department of Health, Education and Welfare was tasked to develop a response to increasing political pressure to meet the needs of the young and the aged.[26] This was not easy because, from the 1930s onward, increasing optimism about the power of the state to ensure the health and wellbeing of the American nation had been organized around concerted response to specific diseases. Since the health of the young and the aged, as distinct subsets of the population, was not accommodated easily within this structure, the Department of Health, Education and Welfare proposed to relocate the various subsections of the specialist, constitutive institutes of National Institutes of Health that were dedicated to either the young or the elderly into a new and especially dedicated institutional structure. This led to the establishment of the National Institute of Child Health and Human Development, which was given the remit to focus 'on the complex health problems and requirements of the whole person rather than on any one disease or part of the body'.[27] Given the historical antecedents of paediatric research, always more influential than gerontology, when the new institution set out to define its objects of inquiry and preferred modes of investigation, the longitudinal approach, that is to say, the temporally extended, serial observation of individuals residing in a particular community, was quickly identified as one of the key areas of attention. Shock, a leading expert in the field, took advantage of the situation. Just a few years earlier, Kennedy's predecessor, Dwight Eisenhower, had established a Federal Council on Aging. Shock had translated the individualist and voluntarist thrust of Eisenhower's initiative into a call to recognize 'older persons as individuals, not as a class, and their wide differences in needs, desires and capacities'.[28] Understanding the ageing process and designing institutions for the elderly required a focus on the individual, and to capture this unit of analysis and intervention, Shock proposed that 'we should … collect observations on the same individual throughout his [sic] life span'.[29] Importantly, like many paediatricians in relation to childhood, Shock was convinced that problems usually associated with ageing were more the result of inadequate mechanisms of social integration than the product of ageing itself, but this made the data he had been collecting with the assistance of the Baltimore Department of Public Welfare suspect because his human subjects were recruited from an institutionalized population and thus not 'free from clinical signs or history of diseases'.[30] These subjects, Shock and many others maintained, presented a distorted picture of the ageing process that could not be translated into the envisioned reform of public policy. When William Peters offered to recruit his neighbours in Scientists' Cliff, an exclusive residential community on the shores of Chesapeake Bay, Shock took up the opportunity to balance the picture of the ageing process. The importance of this encounter was summarized by the secretary of the Department of Health, Education and Welfare when described the work of the gerontology branch in the following terms:

> Until recently (October 1957), the Gerontology Branch, in its clinical research, used individuals of different ages (20–90 years) as research subjects. These people were medically indigent, admitted either to the Baltimore City Hospitals or the Old People's Home. Thus they did not represent the great majority of the aged who live normal lives in their own communities. Dr. Shock thus felt the need for more balanced research – another side of the aging coin. This need gave birth to a unique program of research on the aging process, now in its second year, the Scientists' Cliff Project.[31]

Shock was ambivalent about naming the project after a particular location because this represented a geographical limitation to a programme which he hoped to expand beyond the confines of Scientists' Cliff, but he could not possibly object to the accompanying political recognition that there was 'another side of the aging coin' which could only be captured by studying individuals 'who live normal lives in their communities'. Reforms of health and welfare, as Shock himself put it two years later, were to come 'from studies in which the laboratory is, in effect, the community'.[32] The Baltimore Longitudinal Study of Aging was thus born, and as the exemplar of a programme within gerontology which incorporated demographic and physiological approaches and which embodied how best to solve the problems presented by an ageing population. Not surprisingly, Shock played a pivotal role in the formative endeavours of National Institute of Child Health and Human Development.

If the Baltimore Longitudinal Study of Aging thus provides a further illustration of the role that biological standards played in the alignment of biological research, medicine and American society which took place during the second half of the twentieth century, there was however something unsettling about the Baltimore Longitudinal Study of Aging. The use of serial observation of individuals aimed to clarify the processes involved in the onset of age-related, chronic illnesses within a population, but this was at odds with the ambition to establish simultaneously the parameters of 'physiological age' or normal ageing processes, processes free of all age-related illness. Such uncertainty about what exactly was the object of study may explain Shock's very early interest in Leonard Hayflick's work, the two being in personal contact from 1964, presumably because Hayflick's work offered the scope to create a new experimental system with which to examine the process of normal ageing at the cellular level and so renew Edmund Cowdry's formative understanding that ageing was a properly physiological phenomenon.

As discussed in the previous chapter, during the CIBA Foundation Colloquium on Ageing, Shock had sought to broker a compromise between Cowdry on the one hand, and Alex Comfort and Peter Medawar on the other. In the course of this endeavour, Shock had observed that: 'if you knew what it was that caused a cell to lose its ability to maintain concentration gradients, maintain its metabolic processes, you would be a long way toward understanding the ageing process'. Unfortunately, Shock also observed that:

> The techniques that we have for investigating single cells are very meagre. Dr. Cowdry feels that if you take a cell out of its tissue it is no longer a cell. If we accept this position we are limited to unicellular organisms for study, but unfortunately most of these species simply divide and form two new cells so that 'ageing' fails to occur.[33]

As will be discussed in a later chapter, the notion that ageing is a feature of multicellular organisms alone is important not just to biological, but also philosophical understanding of death. In the meantime, Hayflick and Paul Moorehead claimed instead that somatic cell lines did not reproduce indefinitely and that the cell's life cycle was best understood as divided into three phases, during the last of which the cells began to lose their defining cytogenetic characteristics and to degenerate, so much so, they wrote, that 'entry into Phase III may bear directly upon problems of ageing, or more precisely, "senescence"'.[34] Furthermore, by 1965, Hayflick was claiming that 'the basic step in the Phase III phenomenon is an accumulation of "hits" or errors in DNA replication which inactivates part of the genome'.[35] Here was the answer to the questions Shock has posed ten years earlier, so it is not at all surprising that he should have invited Hayflick to contribute to *Growing Old: Perspectives in Experimental Gerontology* (1966). a volume to which the major experimental biologists of the day contributed. As Shock put it in the preface to the volume:

> The goal of the book is to present an overview of ideas in the field of experimental gerontology, with emphasis on 'ideas' and 'experiments'. Therefore it represents more a collection of essays than a detailed summary of all the factual data on specific problem areas ... The selection of contributors to the volume was a difficult task since every investigator in experimental gerontology should have been included. Had this volume been assembled in 1955 instead of 1965 the task would have been simple, since the number was so few that all could have been included. However, in 1965, the number of competent investigators working on various aspects of experimental gerontology was so large that arbitrary selections had to be made.[36]

This said, Shock did not explain what criteria he adopted to select his contributors and what is most striking about the volume is the uncertainty about what exactly Hayflick's findings contributed to understanding the biology of ageing and death. This is thoroughly consistent with the very mixed responses of that small number of researchers beside Shock, who also took note of Hayflick and Moorehead's findings. For example, in 1962, John Maynard Smith, one of Medawar and Comfort's colleagues at University College London, suggested in a review of the literature that the observed mortality of the cell lines seemed to be an effect of subdivision, but he admitted that the 'mechanisms underlying ... progressive and more or less synchronized changes in clones are not understood'.[37] A few years later, Bernard Strehler, author of *Time, Cells and Ageing* (1962), and his colleagues in the gerontology branch offered further experimental evidence to support the notion of declining viability of cell lines. They regarded the decline as systemic, however, and so called into question not just the division of the cell's life cycle into three distinct phases, but also the speculative explanation that the cell's limited potential to subdivide was determined by some still indeterminate nucleic mechanism.[38] Clearly, Hayflick's discovery was regarded as significant, but its importance may have been related, at least initially, to the experimental system which it brought into gerontology, and much needed to be clarified before the intimation that 'entry into Phase III may bear directly upon problems of ageing, or more precisely, "senescence"' could be translated into any epistemic and institutional redefinition of the field.

The multiple meanings of the normal and the pathological

Reaffirming the cautions issued about any misunderstanding of the biographical method *Biopolitics and the Philosophy of Death* would seem to employ, the uncertainty surrounding the significance of Leonard Hayflick's contribution to understanding the biology of ageing partook in the uncertainty besetting American gerontology more generally.

While the integration of gerontology in the programme of the National Institute of Child Health and Human Development marked federal recognition of the importance accorded to the aged and their needs, in actual fact the programme focused primarily on the other end of spectrum, children and their needs. Consequently, as Betty Lockett observes, gerontologists spent most of the 1960s lobbying on behalf of a federal programme that would better 'coordinate research on the biological origins of aging'.[39] As Andrew Achenbaum argues, however, the development of gerontology continued to be shaped by the historical association of the ageing and the Veterans Administration. As increasing numbers of veterans of the world wars reached old age, calls to establish a more proactive national programme of research into the causes of the diseases commonly associated with old age began to gain substantial political support.[40] At the same time, this reinforced an increasingly divisive equation of ageing and illness such that, when the establishment of such a programme finally moved onto the national political agenda during the second half of the 1960s, the drafting of legislation to create the desired federal programme or institute proved politically difficult, even including a presidential veto.[41] In other words, there were opportunities for development, but there was also considerable uncertainty about how these were best pursued. In this context, Hayflick's growing stature among federal regulators was, for gerontologists, an important asset to the development of research on the biology of ageing.[42] If such importance was to be mobilized to the advantage of the discipline, however, it was important to resolve the uncertainties surrounding Hayflick's contribution to the understanding of ageing as biological phenomenon. This process of integration offers not only a splendid example of the ambiguities of the molecular perspective, but also exemplifies, to paraphrase Luc Boltanski and Laurent Thévenot (1991), how the work of justification involved in the coordination of heterogeneous activities is a double-edged sword.

Established experimental gerontologists such as Bernard Strehler shared much of Hayflick's understanding of the future of gerontology, Strehler having argued as early as in 1967 that:

> The understanding of the nature of the aging process at the molecular, cellular, and organismic level is one of the last virgin territories in biology; although the last decade has seen a number of significant incursions into this wasteland of understanding, it is curious that a property of life that is so universal as aging and so potent in its implications for man as a perceiving and reasoning animal has been relegated *de facto* to the bottom of the ... list.[43]

Strehler regarded Hayflick's as one of these 'significant incursions'. True to the long-standing understanding of ageing as a complex phenomenon, however, Strehler also regarded ageing as a systemic effect and was not persuaded by any notion of a single, universal mechanism, which is what he understood Hayflick to have proposed. The challenge, Strehler argued, was to capture the process of 'normal ageing', which he regarded as the progressive loss of functional integration, and to do so by offering causal explanations grounded in the details of molecular dynamics. He thus speculated that senescence was the cumulative effect of copying errors and other kinds of genetic damage over the individual's lifetime, so offering a translation into molecular terms of the well-established understanding of ageing as a consequence of wear and tear.

From Hayflick's perspective, Strehler's understanding, while welcome when viewed from a methodological point of view, was also unwelcome because it undermined the qualitative distinction implied by his tripartite division of the cell's life, which served to secure the boundaries between normal and abnormal cellular reproduction, and so threatened the coordination of gerontology with oncology and virology. This is perhaps how best to understand Hayflick's increasing insistence that there had to be something like a genetic mechanism which determined the onset of the unpredictably diverse phenomena associated with ageing, or, as he put it, 'the basic processes of aging ... may result from the deterioration of the genetic program that orchestrates the development of cells'.[44] By the late 1960s, Hayflick was setting his doctoral students to work on the putative genetic programme and, almost as if he were responding to Strehler's own mobilization of August Weismann's evolutionary reflections upon the biology of ageing, he also began to move closer to Alex Comfort's own explanation of ageing as the result of 'programmed senescence'.[45] As discussed in the last chapter, Comfort had proposed that natural selection operated most forcefully on those phases of the life cycle that were related to reproduction, so that the expression of any deleterious mutations in these phases would be targeted more strongly than their expression after reproduction. This, again according to Comfort, led to an accumulation of deleterious genes whose expression occurred in the later phases of the life cycle, eventually resulting in the post-reproductive weakening of the organism commonly known as ageing. Importantly, Comfort, as noted earlier, concluded that ageing 'should therefore be treated as a unity of effects rather as a unity of causes'.[46] This resonated greatly with Hayflick's notion that, in the third of his three phases of the cell's life cycle, growth continued, but was increasingly uncontrolled and disorderly, so that the cells increasingly resembled cancerous ones. Strikingly, in his Kleemeier Lecture, Hayflick discussed the menopause and the greying of hair as conditions 'not ordinarily thought of as states of disease but clearly are regarded as early signs of the normal processes of biological aging'. He explained that:

I have chosen these two examples because they occur subsequent to that period in the human life cycle when reproductive capabilities are most vigorously expressed. For it seems clear that from the standpoints both of evolution and survival of the species, age-related biological decrements occur after reproductive maturation. I am not persuaded at all by the frequently stated notion that aging begins at conception. In the sense that the genetic program for age changes may be set at that time, there is no contest, but it would be more accurate to say that biological manifestations of age changes generally occur subsequent to species members reaching reproductive maturity.[47]

From this evolutionary perspective, however, seeking to separate physiological changes involved in pathological states from 'the basic biological changes that are a part of the aging process' was questionable, to say the least. As Hayflick put the matter on the same occasion:

> One is forced to conclude that if all disease related causes of death were to be resolved, then the aging processes would present some clear physical manifestations well in advance of death itself. The challenge, of course, is to separate disease-related changes from the basic biological changes that are a part of the aging process. Since fundamental aging processes most certainly contribute to or allow for the expression of pathology, then the two concepts may be so closely intertwined as to make any clear distinctions a futile exercise in semantics.[48]

In other words, Hayflick's programme for gerontological research held out the promise of aligning gerontology with the ongoing epistemological and institutional transformation of American biology and medicine, but the clarification of its claims and assumptions also threatened gerontologists' shared, but uneasy and unstable commitment to distinguishing between normal and pathological ageing, between good and bad ways of growing old.

For Hayflick, the consequences of this unresolved tension at the very heart of gerontology became most evident upon the establishment of the National Institute of Aging.

Ageing and the politics of anguish

In 1976, just two years after his critical declarations on the occasion of the Kleemeier Lecture, Leonard Hayflick was a candidate for appointment as first director of the newly established National Institute of Aging. The first director was not Hayflick, however, but Robert Butler, author of the best-selling and prize-winning book *Why Survive? Growing Old in America* (1975).[49] Much has been said about Hayflick's fall from grace as he became involved in a protracted and very visible legal litigation with the National Institutes of Health over the ownership of the WI-38 cell line, which Hayflick had developed while still at the Wistar Institute, with funding by the National Institutes of Health, and then multiplied on a commercial basis for sale to the pharmaceutical company Merck & Co.[50] Hayflick's ambition to totally reconfigure the organization of gerontological research, breaking all existing conceptual and institutional alignments between laboratory, clinic and community, seems however a far more credible reason for its limited success. If the development of cell lines that could stand for the human organism, and yet were wholly disconnected from their source of origin, represents a historically important link between Hayflick and the contemporary, increasing alienation of life and the human organism, the course of human life in later years continued to be critically important to the evolution of the National Institute of Aging.[51]

One of the greatest challenges confronting Butler upon his appointment as director of the National Institute of Aging was the lack of research capacity and limited interest in the aged among much of the American public. When Butler suggested in one of his first public reports on gerontological research within the new organization which he was called to lead that 'research on aging has shifted from its exclusive disease orientation toward a more comprehensive investigation of the normal, physiological changes with age', the evocation of normality was related primarily to the need to rearticulate how the public viewed older people's role in society, rather than to the needs of clinicians working with older people and used to distinguishing between normal and pathological states.[52] Despite such attempts to start with a positive and affirmative vision of ageing, however, the National Institute of Aging struggled to secure any steady stream of resources, and this situation only began to change with the emergence of what Butler himself called the 'health politics of anguish', an alliance of activists, clinicians and politicians who called public attention to the psychic abandonment experienced by the aged. Butler himself had set out the following challenge in *Why Survive?*:

> Can we find ways to create diversification within ourselves? Excessive or exaggerated identity seems clearly to be an obstacle to continued growth and development throughout life and to appreciation of the future. I would go as far as to say that a continuing lifelong concern with one's identity is a sign of good health, and the right to have such concern is one of the important rights of life. Human beings need the freedom to live with change, to invent and reinvent themselves a number of times throughout their lives. By loosening up life we enlarge the gift of life.[53]

Once translated into senile dementia, this concern about the importance of retaining the ability to manage change and preserve a sense of identity in the face of change resulted in the prioritization of research into the causes and treatment of Alzheimer's Disease, particularly through the National Institute of Aging's extramural research programme, which, according to Tiago Moreira (2009), embodied emerging, competition-driven innovation policies that were based on both collaborations between universities and pharmaceutical companies and ideals of rational therapeutic development from bench to bedside. It should be noted furthermore that, despite Butler's consequent disenchantment with the direction taken by the National Institute of Aging, whereby ageing was equated increasingly with disease, and Alzheimer's Disease in particular, his initial programme could be said to have facilitated the change insofar as it rested on the transformation of individual failures to adapt and age 'successfully' from a matter of psychic disposition into a matter of organic pathology which could be defeated, just like 'polio'.[54] With the support thus gained, the National Institute of Aging experienced an extraordinary influx of researchers from other areas of biomedical research. This helped to transform the National Institute of Aging's place in the American polity, but in the form of a highly visible disease-specific programme that eventually accounted for the majority of the Institute's budget. It also caused considerable dissatisfaction among a number of gerontologists and activist who complained that ageing was much more than a matter of increasing senility and eventual death. As Richard Miller, one of the signatories to the transatlantic biogerontological proposal discussed earlier, has noted and plaintively so, 'senators' and voters' parents [die] of specific diseases' and are less likely to fund a general, 'basic' programme of research on ageing.[55] In sum, Alzheimer's Disease firmly established the position of the National Institute of Aging within the political and clinical worlds, but only by reinforcing ever increasingly the equation of ageing and illness.[56] It now is very difficult to think otherwise.

Of molecules and organisms

In 1973, during the annual meeting of the Gerontological Society of America, Leonard Hayflick, proponent of the notion that ageing and mortality were intrinsic properties of the cell, offered what seemed like a new and exciting agenda for the future development of gerontology and better understanding of the biology of ageing and death. The interest shown in Hayflick and his research, which, by Hayflick's own admission, was the product of considerations wholly extraneous to the life history of the ageing organism, was linked to gerontologists' expectation that Hayflick's investigations of cellular reproduction would renew Edmund Cowdry's programme for gerontological research by establishing the physiologist's laboratory as the pre-eminent site where to resolve the challenges confronting a society wherein the aged were increasingly visible and vocal. It was also hoped that it would align gerontology with the growing field of molecular biology and its own ambition to reconfigure the entire domain of biology, epistemic and institutional. At the same time, however, the very alignment with experimental oncology and the development of new biological standards, which helped to secure the institutional stability of Hayflick's iconoclastic claims about the mortality of cell lines, also drove the evolution of these claims toward an understanding of ageing whereby it became difficult to distinguish between ageing and illness. As such, Hayflick's vision also threatened to undermine the bonds between biology, medicine and society that American gerontology had forged from the 1930s onward by insisting upon the importance of distinguishing between normal and pathological states. Such difficulties are of some importance to the argument that *Biopolitics and the Philosophy of Death* is advancing.

As Michel Foucault observed in *The Birth of the Clinic* (1964), the emergence of a distinctively modern understanding of life is inseparable from the fundamental reconfiguration of the relationship between life and death, which Foucault associated with Xavier Bichat's definition of life as 'the sum of all activities that resist death'.[57] For a century and more, this understanding of the relationship between life and death served to define the living organism as an integrated unit and site of 'free and independent life', as Cowdry put it in *Problems of Ageing* (1939).[58] Throughout these years and beyond, the organism was also the undisputed site of negotiation between the arguably incommensurable worlds of biology and medicine, and, as Georges Canguilhem observed in *The Normal and the Pathological* (1966), the distinction between normal and pathological states was critically important to these negotiations.[59] As Canguilhem also observed, however, genetic concepts and thought, always located ambiguously between the molecular and evolutionary modes of inquiry, disrupted this distinction by effectively destabilizing the organism and all the practices associated with the organism.[60] Hayflick's and Alex Comfort's provocative, if not heretical, views on the nature of ageing certainly attest to the truth of this. Hayflick and Comfort contributed signally to the alienation of life and organism such that the latter should be regarded as the secondary product of the dynamics of molecules and populations, but, ultimately, they both also failed to secure a corresponding transformation of gerontological research because their views were at odds with the congruence of clinical and popular belief in the centrality of the ageing organism and in the possibility of exorcizing all understanding of this organism of any reference to decline and death. Robert Butler, tireless critic of all pathologization of old age and champion of the notion that 'human beings need the freedom to live with change, to invent and reinvent themselves a number of times throughout their lives', thus regarded the increasing integration of gerontology within the structures of American biomedical research with great satisfaction, as it indicated that 'research on aging has shifted from its exclusive disease orientation toward a more comprehensive investigation of the normal, physiological changes with age'. Unfortunately, however, these very same ambitions to forge a better understanding of the entire course of life were only fulfilled by equating ageing, senile dementia and Alzheimer's Disease. Securing a coherent and positive, if not affirmative, understanding of life in later years within the expanding domain of biomedical institutions seemed impossible. Unsurprisingly, the relationship between biology and medicine eventually became the principal target of the biogerontological critique of biomedicine mounted by Butler, Hayflick and others from the late 1990s onward. In the meantime it seems that while ageing and death are undeniably subjects of biomedical inquiry, they also seem wholly refractory to, and to defy any organization into the single 'space of representation' which Peter Keating and Alberto Cambrosio evoke in *Biomedical Platforms*, as they seek to revisit Canguilhem's analysis of the relationship between the normal and the pathological, the laboratory and the clinic, and between biology and medicine more generally. There may be, in other words, something very different and destabilizing about ageing and death that seems uniquely capable of highlighting the limits of all intellectual and institutional formations.

CHAPTER 5

Forging the future

The task of the genealogist is to clarify how best to make sense of the present situation for the sake of the future. Playing with Friedrich Nietzsche's and Michel Foucault's reflections on historiographical practice, the genealogist tells 'histories of life' and does so 'for life'.[1] The claim advanced over the preceding three chapters of *Biopolitics and the Philosophy of Death* is that the bounding of life as unfolding between the birth and death of the organism is a historical construction, and that the contours marked out are today uncertain, largely because the biomedical sciences seem to be transforming every aspect of human existence imaginable, yet, at the same time, biomedical understanding of ageing and death is replete with ambiguities and obscurities. This chapter will seek to outline how biogerontologists seek to forge a more determinate future by mobilizing this past and how they do so on two levels, narrative and performative. As already observed, cultural critics are not the only ones to tell stories about the historical development of the biomedical sciences. Those who are directly involved in the development of these sciences also tell stories and their stories matter to the construction of the objects that define these sciences.

In a historical review of 'cytogerontology', the neologism that Leonard Hayflick had coined for the new approach to the study of ageing which he sought to advance and which was discussed at great length in the last chapter, Thomas Kirkwood and Thomas Cremer observed how Hayflick's work on the life course of the cell had 'opened fresh speculation into the possibility that ageing might be due to intrinsic limitations in the life of somatic cells'.[2] Kirkwood and Cremer also suggested, however, that such limitation had to be understood within a wider evolutionary framework than was usually considered at the time of writing because, although geneticists had explained how ageing became manifest at a particular moment in the life of the organism, these same geneticists had left unexplained why organisms were required to age and die. While Kirkwood and Cremer's argument is deceptively similar to Alex Comfort's, it rests on a critical re-evaluation of August Weismann's legacy. As observed earlier, Weismann's distinction between germinal and somatic constituents has proven fundamentally important to the emergence of the genetic theory of heredity and its ambitions to account for all biological phenomena, from the genetic to the evolutionary levels. Kirkwood and Cremer return to this pivotal distinction, drawing attention to the questions it raises about the construction of the organism's trajectory from birth to death. Importantly, the recovery of this distinction and its reinterpretation has proven critically important to the development of the transatlantic critique of contemporary understanding of the biology of ageing, which, as was discussed in the last chapter, equates ageing and disease, particularly, with senile dementia and Alzheimer's Disease. The recovery of this distinction allows biogerontologists to reconfigure those complicated trajectories of biodemographic and biomolecular research outlined in previous chapters and so forge a new agenda for the future development of biomedical understanding of ageing and death. On this alternative configuration, the relationship between the mortality of the organism and the immortality of the genetic lineage should redirect gerontologists' attention toward the mechanisms involved in the repair of any damage to the genetic programme linking the two and to the proper maintenance of these mechanisms. This reconstruction enables biogerontology to integrate the two branches of biomedical inquiry, molecular and demographic, with clinical and social perspectives upon ageing and death at a good age. Thus, as Tiago Moreira has observed rightly, 'scientists' stories are not only a topic in the practical accomplishment of the history and social studies of science, technology and medicine, but also in the accomplishment of facts, artefacts, representations, and bodies themselves'.[3] Just as usefully, Laurent Thévenot has sought to correct such understanding by arguing that, because they do not just mobilize the past, but also put it to work, to configure the future, these scientists are not simply 'practical historians', but also 'critical historians'.[4]

    Critical histories are not always effective, however. After

attending to biogerontologists' genealogical endeavours, this chapter will turn first to Michel Foucault's lectures on security and biopolitical governance, aiming to account for the political resonance of the biogerontological argument about the nature of ageing and its accompanying practices of intervention. At the same time, however, the chapter will also seek to explain why the reception accorded to the biogerontological perspective has in fact proven to be far more ambivalent than might have been expected in light of such resonance. It seems that administrative institutions of modern states such as the United Kingdom, the Department of Work and Pensions in particular, remain wedded to those disciplinary structures and organizations that Foucault first outlined in *Discipline and Punish* (1975). The chapter will then draw to a close by reflecting on these difficulties and how they call for the re-examination of our most fundamental conceptual categories, so creating a bridge to the third part of *Biopolitics and the Philosophy of Death*.

Re-reading Weismann

During the 1990s, the alignment of evolutionary models and genetic research, which Richard Dawkins's *The Selfish Gene* (1976) had popularized very successfully, renewed gerontologists' interest in the evolutionary understanding of ageing. While seeking to explain how the diversity of living forms was best understood by focusing on genes and the dynamics of their reproduction, Dawkins wrote that:

> The question of why we die of old age is a complex one, and the details are beyond the scope of this book. In addition to particular reasons, some more general ones have been proposed. For example, one theory is that senility represents an accumulation of deleterious copying errors and other kinds of gene damage which occur during the individual's lifetime. Another theory, due to Sir Peter Medawar, is a good example of evolutionary thinking in terms of gene selection. Medawar first dismisses traditional arguments such as: 'Old individuals die as an act of altruism to the rest of the species, because if they stayed around

when they were too decrepit to reproduce, they would clutter up the world to no good purpose.' As Medawar points out, this is a circular argument, assuming what it sets out to prove, namely that old animals are too decrepit to reproduce. It is also a naive group selection or species-selection kind of explanation, although that part of it could be rephrased more respectably. Medawar's own theory has a beautiful logic … According to this theory then, senile decay is simply a by-product of the accumulation in the gene pool of late-acting lethal and semi-lethal genes, which have been allowed to slip though the net of natural selection simply because they are late-acting.[5]

Dawkins thus rehearsed the criticism that much thinking about ageing was prey to circular reasoning and then drew attention to the accidental consequences of evolutionary design. He went on to suggest that, as a result of such design, longevity could be extended either by delaying the age of reproduction, so that the expression of the deleterious genes associated with the debilities of old age was also delayed, or by pharmacological interventions designed to neutralize the physiological mechanisms triggering the expression of these same genes. As Dawkins wrote:

One of the good features of this theory is that it leads us to some rather interesting speculations. For instance it follows from it that if we wanted to increase the human life span, there are two general ways in which we could do it. Firstly, we could ban reproduction before a certain age, say forty. After some centuries of this the minimum age limit would be raised to fifty and so on. It is conceivable that human longevity could be pushed up to several centuries by this means. I cannot imagine that anyone would seriously want to institute such a policy. Secondly we could try to 'fool' genes into thinking that the body they are sitting in is younger than it really is. In practice this would mean identifying changes in the internal chemical environment of a body which take place during ageing. Any of these could be the 'cues' which 'turn on' late-acting lethal genes. By stimulating the superficial chemical properties of a young body it might be possible to prevent the turning on of late-acting deleterious genes … What is revolutionary about this idea is that S [the chemical substance differentiating young and old bodies by triggering the action of the late-acting lethal genes] itself is only a 'label' for old age. Any doctor who noticed that high concentration of S tended to lead to death, would probably think of S as a kind of poison, and would rack his brain to find a direct causal link between S and bodily malfunctioning. But in the case of our hypothetical example, he might be wasting his time![6]

As Dawkins himself acknowledged, however, all this was the stuff of science fiction, which Kirkwood explores when bringing *Time of Our Lives* to a close. There can be little doubt, nonetheless, that this grand vision captured very effectively the growing confidence about the potential of genetic research to transform biology and medicine.[7] Gerontology, as Kirkwood attests, was not immune to the process because the vision offered scope for a new articulation of the problems of old age and how best to address them. Kirkwood's work on the regulation of genetic functions and its relationship to the ageing body illustrates this argument, and it does so very tellingly.

Like Leonard Hayflick's, Kirkwood's career began with the investigation of biological standards, though viewed this time from a statistical perspective. During the early 1970s, while working in the statistics section of the newly established National Institute of Biological Standards and Control, something like the British counterpart to the Division of Biologics, Kirkwood began to collaborate with Robin Holliday, who was then responsible for the programme in experimental oncology at the Medical Research Council's National Institute for Medical Research. Holliday had been working on the mechanisms controlling gene expression. Once it had been agreed that cancerous cells were unusual because they could replicate indefinitely, the issue was to understand the processes involved in oncogenetic transformation and 'immortalization'. In the first of a series of papers on the replicative capacities of different cell cultures, Kirkwood and Holliday turned the problem around by proposing that the principal challenge should be to understand why cells were sometimes unable to reproduce indefinitely. As they put it in one of their first co-authored papers:

> In a population of growing cells, there is continual selection for those with the shortest generation time. It would therefore

be predicted that slow growing senescent cells would be eliminated, leaving the healthy ones to replenish the population. The problem therefore is not to explain the immortality of transformed lines, but rather to account for the inability of cellular selection to keep diploid cultures in a healthy growing condition.[8]

They went on to propose a mathematical model on the assumption that the diversity could be understood in terms of cellular 'commitment' such that genetic programmes of cell development, metabolism and reproduction were either active, resulting in irreversible progress towards death, or dormant, resulting in immortality. Following Leslie Orgel (1963, 1973), the mortality of committed cells was due simply to the accumulation of random errors involved in the process of translating genetic information into the proteins required for cell development, metabolism and reproduction. Crucially, the observed proportion of the two types depended on the size of the population, the number of generations to commitment and the probability of commitment. The model engaged with Hayflick's arguments about the mortality of somatic cell lines by positing that the observed mortality was an artefact of the usual practice of discarding all but a few cultured cells during each generational passage, a practice whereby any rarer, immortal cells were at greater risk of extinction simply because they were numerically fewer, rather than intrinsically different. As Kirkwood and Holliday put it:

> If our model has validity, we can conclude that diploid fibroblasts may in fact be immortal, contrary to the numerous experimental observations which have been published. We predict, however, that such immortality will never be seen in laboratory populations, but only in those of enormous size. If a way could be found to separate uncommitted cells from committed ones, then the population could be kept growing indefinitely.[9]

In sum, Holliday and Kirkwood's explanation of the difference between cellular mortality and immortality was fundamentally statistical insofar as the vital processes upon which they focused were regarded as subject to the vagaries of sampling, to what Michel Foucault labelled the vagaries of the 'aleatory field'.[10]

At about the same time, Kirkwood extended the argument about developmental commitment beyond the confines of experimental oncology. Originally, Kirkwood and Holliday had professed no interest in why some cells were mortal, but, from an evolutionary perspective, the overproduction of somatic cells could be regarded as an adaptation that served to diminish the probability that germinal cells might be lost to external, accidental causes of mortality. Contributing signally to the resurgence of evolutionary biology, of which Richard Dawkins's *The Selfish Gene* was but the most visible and most widely acclaimed manifestation, Kirkwood went on to argue that the organism should then be understood as the product of a process of balancing investments in the somatic body, to enhance the chances of successful reproduction, and the cost of these investments to the continuity of the genetic line.[11] The argument responded to Kirkwood's initial observation that although population geneticists had explained how ageing became manifest at a particular moment in the life course of the organism, they had left unexplained why organisms were required to age and die. As Kirkwood and Holliday put it during a meeting that was to prove fundamentally important in the history of modern evolutionary biology:

> The importance of the work of [Peter] Medawar, [George C.] Williams and [William D.] Hamilton is the clear elucidation of the effect of age on the force of natural selection. This cannot fail to have relevance to the evolution of ageing since it shows how selective control over the later portions of the lifespan must progressively tail away. There is, however, a serious logical objection to explaining the origin of ageing by the theory of accumulating late-expressed deleterious genes ... The objection is that the concept of 'late expression' itself implies the prior existence of adult age-related physiological processes. An organism that does not age is, in effect, in a steady state, where the physiology of chronologically young and old animals is the same. In a situation like this it is difficult, if not impossible, to see how genes could measure the passage of time and have early or late expression. However, if ageing already existed within a population, there would certainly be the potential for precession or recession of good and bad genes since the ageing process itself could define the time scale. In this case, however, the accumulation of deleterious genes at the end of the lifespan would be a reinforcing consequence, not the cause,

of ageing. The same objection applies equally to the pleiotropic gene hypothesis, except where the bad gene effects are simply the cumulative result of the continued expression of a gene which has outlasted its useful function. In general, the risk of circularity in the Medawar–Williams argument needs more careful attention than it has hitherto received.[12]

Basically, Kirkwood and Holliday claimed that the leading evolutionary explanations of the day could offer no properly causal explanation for the increasing toll of natural selection over time because they assumed at the very outset the existence of physiological processes related to the organism's chronological age, the very phenomenon which they set out to explain. The way forward, according to Kirkwood and Holliday, was instead to re-examine the most basic assumptions about the relationship between life, death and the organism.

If, as Dawkins maintained, August Weismann was to provide the foundations of a proper appreciation of the importance of genes and the dynamics of their reproduction to understanding evolutionary processes, Kirkwood and Thomas Cremer (1982) proposed a narrative in which Weismann became a founding, but much misunderstood, figure.[13] The chief source of such misunderstanding, according to Kirkwood and Cremer, is that Weismann wrote not one, but two essays on the evolution of mortality, 'The Duration of Life' (1881) and 'Life and Death' (1883). According to Kirkwood and Cremer's reading of the later essay, Weismann maintained that the emergence of the multicellular organism composed of two different types of cells, somatic and germinal, served to better ensure the continuity of any genetic lineage against accidental causes of mortality. It is this more general and basic understanding that, Kirkwood and Cremer claim, underpins Weismann's following statement:

> Probably at first the somatic cells were not more numerous than the reproductive cells, and while this was the case, the phenomenon of death was inconspicuous, for that which dies was very small. But as the somatic cells relatively increased, the body became of more importance as compared with the reproductive cells, until death seems to affect the whole individual, as in the higher animals, from which our ideas upon the subject are

derived. In reality, however, only one part succumbs to natural death, but it is a part which in size far surpasses that which remains and is immortal, the reproductive cells.[14]

In other words, the difference between mortality and immortality is a matter of numbers and, most importantly, the idea that death is a fundamentally important biological phenomenon is an artefact of the way in which the reflective human subject, as individuated, embodied and mortal organism, cannot but prioritize the organism. The task then is to take seriously Comfort's observation that 'gerontology is an entity which only comes into existence to describe a process human beings don't like', asking, as it were, why the process of ageing is regarded so negatively and what might be done about it within the constraints imposed by molecular and evolutionary dynamics.[15] Such juxtaposition of the organism and the dynamics of its constitutive parts, molecules and populations, has proven critically important to the reconsideration of the business of gerontology from the 1990s onward. Before embarking upon the exploration of such reconsideration, however, it is important to understand more fully the resonance of statistical arguments and their material consolidation.

Securing the population

In many ways, Thomas Kirkwood's thinking about life, death and the organism illustrates powerfully Georges Canguilhem's thesis about the increasing reduction of all biological differentiation to a matter of quantitative variation. Here, all essential distinction between mortality and immortality is eliminated, and any remaining differentiation is reduced to a question of statistical aggregation and the contingencies of sampling. As intimated earlier and with respect to the growing importance of the aleatory field, the forms of reasoning involved are related inescapably to the modern transformation of governance that is the subject of the two series of lectures that Michel Foucault gave at the Collège de France between 1975 and 1978.[16]

The first set of lectures, delivered between 1975 and 1976, was dedicated to the shift from sovereign to biopolitical deployments of power. The second, delivered between 1977 and 1978, charts the diverse governmental rationalities that have shaped the emergence of the modern state. From the eighteenth century onwards, Foucault argues, security and the population take centre stage. Foucault spends much time describing how the modern technology of security differs from either the deployment of law or the implementation of disciplinary powers. Whereas law is a negative power, intent as it is on prohibiting particular activities, and discipline is a positive power that seeks to oblige by prescribing alternative activities, security is neither a negative nor a positive power. Instead, according to Foucault, security lets events take their course and then reacts to them in order to extract some governmental advantage. The techniques used during different historical periods to respond to disease illustrate very usefully the differences between the three modes. While the response to leprosy during the fourteenth and fifteenth centuries had been juridical, resting primarily on exclusion of the leper from the community, the disciplinary response to the plague in the sixteenth and seventeenth centuries had rested on quarantine. The response to smallpox, from the eighteenth century onwards, was altogether different. Rather than deploying techniques of exclusion or quarantine, the focus of medical intervention came to rest upon the determination of probabilities and statistical averages. As Foucault explains, the fundamental problem involved knowing the following:

> How many people are infected with smallpox, at what age, with what effects, with what mortality rate, lesions or after-effects, the risks of inoculation, the probability of an individual dying or being infected by smallpox despite inoculation, and the statistical effects on the population in general.[17]

The aim of medical intervention under security was neither to prevent contact between the sick and the healthy, nor to treat the disease in all patients, but to use the knowledge above to establish the 'normal distribution' of smallpox across a population that included both the sick and the healthy. On the basis of this distribution, Foucault claims, medical professionals were able to identify and reduce the most extreme deviations from the statistical norm. Importantly, like the techniques that had previously been developed to deal with leprosy and the plague, security was not limited to medicine. From the eighteenth century onwards, security operated in a variety of other domains. Foucault discusses, for example, how the mode of security came to organize urban life. It did not attempt to plan everything in advance, down to the last detail, but rather it sought to estimate the volume of people and goods likely to circulate through the city and then sought to plan the construction of houses, streets and districts on the basis of these possible events. Similarly, security dealt with the provision of food not by attempting to prevent scarcity in advance, but instead by forecasting the possible effects of scarcity before they occurred and by then dealing with them as and when they arose. In other words, security allowed things to 'take their course' and then managed the result of a process whose random effects could be minimized or cancelled out by the prior calculus of probabilities. In contrast to legal prohibition and disciplinary regulation, security then was concerned with the management of the inevitable processes that took place on the largest scale imaginable, which is to say that it was concerned with processes at the level of the population. Finally, to these transformations corresponded an equal one, from natural history to biology. Revisiting issues discussed previously, in *The Order of Things* (1966), Foucault explained:

> Basically, as you know, the essential role and function of natural history was to determine the classificatory characteristics of living beings that would enable them to be distributed to this or that case of the table. In the eighteenth and the beginning of the nineteenth century, a whole series of transformations take place that take us from the identification of the classificatory characteristics to the internal organization of the organism and then from the organism in its anatomical-functional coherence to the constitutive or regulatory relationships with the milieu in which it lives. Roughly speaking, this is the Lamarck-Cuvier problem, to which Cuvier provides the solution, in which the principles of rationality are found in Cuvier. Finally in the transition from Cuvier to Darwin, from the milieu of life, in its constitutive relationship to the organism, we pass to the population that Darwin succeeded in showing was, in fact, the element through which the milieu produces its effect on the organism. To think about the relationship between the milieu and the organism, Lamarck resorted to something like the idea of the organism being acted on directly and shaped by the milieu. Cuvier

resorted to what appear to be more mythological things – like catastrophes, God's creative acts, and so on – but which actually organized the field of rationality much more carefully. Darwin found the population was the medium between the milieu and the organism, with all the specific effect of population: mutations, eliminations, and so forth. So in the analysis of living beings is the problematization of the population that makes possible the transition from natural history to biology. We should look for the turning point between natural history and biology on the side of population.[18]

In other words, the introduction of the population separates life and the organism, lending material substance to the statistical form of reasoning installed by the governmental mode of security. In this context, which some other historians have described very aptly as 'the empire of chance', the death of any given individual organism, its 'elimination', is to be regarded as no more than a random event and of no fundamental is significance in the history of life.[19]

As August Weismann could be said to have recognized, however, the organism remains nonetheless important to the bonding of the population into a biologically meaningful entity. Foucault discusses the resulting complication in the earlier set of lectures. Rehearsing themes broached at the same time in *The History of Sexuality* (1976), Foucault also told his audience that:

> Power, which used to have sovereignty as its modality or organizing schema, found itself unable to govern the economic and political body of a society that was undergoing both a demographic explosion and industrialization. So much so that far too many things were escaping the old mechanism of the power of sovereignty, both at the top and the bottom, both at the level of detail and at the mass level. A first adjustment was made to take care of the details. Discipline had meant adjusting power mechanisms to the individual body by using surveillance and training. That, of course, was the easier and more convenient thing to adjust. That is why it was the first to be introduced – as early as the seventeenth century, or the beginning of the eighteenth – at a local level, in intuitive, empirical, and fragmented forms, and in the restricted framework of institutions such as schools, hospitals, barracks, workshops, and so on. And then at

the end of the eighteenth century, you have a second adjustment; the mechanisms are adjusted to the phenomena of population, to the biological or biosociological processes characteristic of human masses. This adjustment was obviously much more difficult to make because it implied complex systems of coordination and centralization.

So we have two series: the body-organism-discipline-institutions series and the population-biological processes-regulatory mechanisms-State.[20]

In other words, governing growing, industrial societies is not easy, and multiple configurations of power and knowledge are therefore required. Then, Foucault asks pointedly, why 'did sexuality become a field of vital strategic importance in the nineteenth century'? Foucault's equally pointed answer, as observed earlier, was:

I think that sexuality was important for a whole host of reasons, and for these reasons in particular. On the one hand, sexuality, being an eminently corporeal mode of behaviour, is a matter for individualizing disciplinary controls that take the form of permanent surveillance (and the famous controls that were, from the late eighteenth to the twentieth century, placed both at home and at school on children who masturbated represent precisely this aspect of the disciplinary control of sexuality). But because it also has procreative effects, sexuality is also inscribed, takes effect, in broad biological processes that concern not the bodies of individuals but the element, the multiple unity of the population. Sexuality exists at the point where body and population meet. And so it is a matter for discipline, but also a matter for regularization.

It is, I think, the privileged position it occupies between organism and population, between the body and general phenomena, that explains the extreme emphasis placed upon sexuality in the nineteenth century.[21]

In sum, sexuality is the site where the coordination of organism and population takes shape, so linking their correlative epistemic structures and juridico-political institutions. As Foucault intimates, but never spells out in the pages following these reflections, genetics, poised as it is between two modes of analysis, physiological and statistical, is the inheritor to the challenge posed by this problem of coordination.[22] In their different ways, Paul Rabinow and Nikolas Rose have sought to chart how the biomolecular sciences have mediated the increasing integration of the 'body-organism-discipline-institutions' and 'population-biological processes-regulatory mechanisms-State' series. As was observed in first chapter, however, the reconfiguration of gerontology, which has taken shape from the 1990s onward, suggests that such integration is beset by considerable difficulties.

The business of biogerontology

As discussed in the last chapter, during the 1980s and 1990s, the National Institute of Aging's programme of research superseded Leonard Hayflick and Robert Butler's vision of research on ageing and its future development. It advanced a programme that could so powerfully align public, policymakers and clinicians that Alzheimer's Disease and the problems posed by an 'ageing society' became virtually synonymous and so displaced any plans to locate research into the causes of ageing at the heart of the biomedical enterprise. If Hayflick and Butler felt that an opportunity had been missed, however, this situation also created the conditions for an unlikely alliance between programmes aiming to distinguish between normal and pathological ageing on the one hand, and investigations of ageing at the genetic level on the other hand. While one of the features of the alliance has been the return to the evolutionary explanation of ageing, its mainstay has been the transformation of its alternative proposal for the organization of research on ageing into a full-blown critique of biomedicine more generally, a critique that denounces the existing alignment of medicine, biology and political institutions as wholly misguided. The evolution of this complex alliance is illustrated by the further development and responses to Thomas Kirkwood's work.

As discussed above, Kirkwood rearticulated evolutionary explanations of ageing by combining molecular analyses and statistical calculations to advance the notion that the organism should be understood as the product of a process involving the balancing of investments of resources in the somatic body, to enhance the chances of successful reproduction of the germinal line, and the cost of these investments to the continuity of this same line. In a comprehensive review for *Nature*, Kirkwood and Steven Austad summarized the implications of this understanding in the following terms:

> An important corollary of the prediction (and emerging evidence) that key genes regulating rate of ageing are those that control somatic maintenance and repair, is that at the level of the individual there is considerable scope for the action of stochastic chance. Not only will individual cells within tissues experience different random accumulations of faults, but there may also be important stochastic variations in developmental processes resulting, for example, in different numbers of cells being formed in key organs, such as the hippocampus. Variation in initial cell number and damage rate will in turn affect the time taken before a threshold for dysfunction is crossed during the progressive neuro-degeneration that occurs later in life. This means that genetically identical individuals, maintained in uniform environments, may nevertheless exhibit considerable variation in aspects of the senescent phenotype, as has been frequently observed in ageing studies on inbred laboratory organisms. Heterogeneity in the senescent phenotype, arising from intrinsic stochasticity as well as genetic and environmental variations, may also help to explain the intriguing phenomenon that in several species age-specific mortality rate eventually slows its rate of increase and may even decline. Heterogeneity explains such an effect if we assume that the frailer individuals die first, leaving a residual population that, as time goes by, represents a dwindling sub-population made up of those individuals that were always the most resilient.[23]

Basically, the claim is that even genetically identical twins, whether raised under identical conditions or not, age at different rates, and that exploration of the difference allows greater understanding of the mechanisms involved in the protection of the germinal lines against the accidents of history, against the very aleatory phenomena to which Michel Foucault once drew attention. On this understanding of ageing, attention is directed toward the molecular mechanisms involved in the preservation of genomic integrity, and the business of gerontology then becomes one dedicated to enhancing the ability of the individual to imitate the immortal germinal line. As Kirkwood and Austad summarized their argument, the attention of all those interested in ageing should be focused upon 'the evolved capacity of somatic cells to carry out effective maintenance and repair'.[24] Although immortality itself has to be regarded as denied irretrievably by the evolutionary history of the human species, the hope is that this redefinition will at least result in maximizing the biological functionality of the individual up to the inevitable moment of death. As Kirkwood puts the argument in the last lines of *Time of Our Lives*:

> Freedom makes us individually responsible for our choices and our actions. Is this why we so readily drug ourselves into inactivity with low-demand time-fillers when we could do so much? Let us be truly alive, so that when old age finally robs us of our vitality, we may feel that the time of our lives was well spent.[25]

Importantly, when viewed from this affirmative perspective, gerontology ceases to be a field of clinical specialization concerned with the diseases of a distinct subset of the population, the elderly, as these diseases are rearticulated as unfolding temporally from antecedent exposure to external risk factors and along biomolecular pathways. The evolution of such understanding is inseparable from Kirkwood's encounter with the contemporary equation of ageing and Alzheimer's Disease.

As noted earlier, with respect to the organization of gerontological research in the United Kingdom, the National Health Service and the Medical Research Council did not greatly support the development of gerontology as a distinct field of biomedical research and clinical specialization, but, during the 1960s, they lent considerable backing to a programme of research on senile dementia at the University of Newcastle and Newcastle General Hospital. As Duncan Wilson, Robert Katzman and Katherine Bick have observed, this programme has proven fundamentally important to the evolution of contemporary understanding of Alzheimer's Disease.[26] Martin Roth, Gary Blessed and Bernard Tomlinson contributed greatly to the revival of biomedical interest in senile dementia by claiming to have established a significant correlation between the scores that elderly subjects, within and without institutional settings, obtained in cognitive tests and the number of neuronal plaques found in these subjects' brains *postmortem*. Significantly, Roth, a clinical psychiatrist, drew attention to the high incidence of dementia among the elderly still living in the community and argued that the robustness of the correlation was potentially very important to the future, rational allocation of resources within the Department of Health and Social Services, between clinical centres and provision of care for the elderly within the community. While Roth's claims about the incidence of dementia among the elderly were relatively uncontroversial, the Department of Health and Social Services was not as persuaded by the robustness of the scale that Roth, Blessed and Tomlinson had developed and by the accompanying claim that this scale allowed them to track the development of a discrete pathological entity, from early signs to total loss of all capacities. Consequently, over the 1970s, the three endeavoured to bring together ever more sophisticated psychometric techniques with developments in electron microscopy that helped to redescribe the putative neuropathological features of dementia at the ultrastructural level. The expanding programme of research was also boosted by Elaine and Robert Perry's simultaneous proposal that the cognitive deficiencies thus measured were not only correlated with these ultrastructural features, but were also related to lowered levels of the neurotransmitter acetylcholine, so holding out the prospect of developing a biomarker for the development of Alzheimer's Disease *in vivo*. Given the evolution of gerontological research in the United States, this emerging understanding positioned the University of Newcastle as an internationally recognized, leading centre of such research, but doubts persisted because, once again, there was no consensus about what constituted the normal course of cognitive capacity and the role of social dynamics in shaping such capacity. In Newcastle, the difficulties involved in coordinating the increasingly multidisciplinary endeavour that was required to answer these doubts, spread as it was across different institutional settings, resulted in the establishment of an institute dedicated to the study of health and illness in the elderly, the Institute for the Health of the Elderly. Kirkwood joined this institution in 1999, moving from the University of Manchester, where he had held the first chair in biological gerontology anywhere in the United Kingdom.

Kirkwood's distinctive understanding of ageing could be mobilized very usefully to argue for the integration of the clinical, biomolecular and biodemographic perspectives, so transforming the position of gerontology within the political economy of the biomedical sciences. As Kirkwood put it in the course of a symposium on 'Mild Cognitive Impairment', a widely heralded, but also contested precursor to the onset and full expression of Alzheimer's Disease:

> Although there have been unquestionable benefits from the labeling of neurodegenerative conditions such as AD [Alzheimer's Disease] as distinct clinical entities, not least the removal of some of the stigma previously associated with senile dementia, the biology of aging suggests that too rigid an approach to classification is of doubtful validity and may even get in the way of understanding what is really going on. The debate about MCI [Mild Cognitive Impairment] not only illustrates the problems with trying to classify what in truth may be unclassifiable, but also may help us to see more clearly the importance of focusing on underlying causes and their effects. When we know more about these underlying mechanisms, we will be better able to tell what the signs and symptoms of early cognitive impairment really mean and what might best be done to intervene.[27]

Following the more complex understanding of ageing that Kirkwood seeks to advance, Alzheimer's Disease must be viewed as part of a much wider set of 'degenerative diseases' that are only connected contingently to the organism's chronological age. Within this configuration, all such conditions might be said to entail ageing, but in so expanding its domain of application the term 'ageing' no longer identifies a distinct biological process of its own kind. Understanding the pathways from the earliest possible manifestations of forgetfulness to the full expression of Alzheimer's Disease then becomes an opportunity to mobilize multiple interventions, medical and behavioural, as part of an integrated programme in public health. The business of biogerontology so defined becomes synonymous with the management of life over its entire course from birth to death, aiming to secure health into advanced age.

Needless to say, this expansive vision of ageing has left its imprint on the Institute for the Health of the Elderly. The Institute had been established with funds from the National Health Service, and its very name signalled the resultant clinical orientation of its work. This became difficult to reconcile with an increasingly centrifugal definition of ageing and the accompanying, equally increasing funding from the Medical Research Council and the Wellcome Trust. Tellingly, the Institute was renamed in 2002, becoming the Institute for Ageing and Health. Two years later, Kirkwood became its director.

Importantly, the ambitious understanding of ageing, which the like of the Institute for Ageing and Health seek to advance, allows biogerontologists such as Kirkwood to differentiate themselves from the anti-ageing movement. As a group of leading biogerontologists, including not only Kirkwood, Butler and Hayflick, as well as Robin Holliday and Linda Partridge, but also Aubrey de Grey, put it in a public statement issued in 2002:

> In the past century, a combination of successful public health campaigns, changes in living environments, and advances in medicine have led to a dramatic increase in human life expectancy. Long lives experienced by unprecedented numbers of people in developed countries are a triumph of human ingenuity. This remarkable achievement has produced economic, political, and societal changes that are both positive and negative. While there is every reason to be optimistic that continuing progress in public health and the biomedical sciences will contribute to even longer and healthier lives in the future, a disturbing and potentially dangerous trend has also emerged in recent years. There has been a resurgence and proliferation of health care providers and entrepreneurs who are promoting anti-aging products and lifestyle changes that they claim will slow, stop, or reverse the processes of aging. Even though in most cases there is little or no scientific basis for these claims, the public is spending vast sums of money on these products and lifestyle changes, some of which may be harmful. Scientists are unwittingly contributing to the proliferation of these pseudo-scientific anti-aging products by failing to participate in the public dialogue about the genuine science of aging research. The purpose of this document is to warn the public against the use of ineffective and potentially harmful anti-aging interventions, and provide a brief but authoritative consensus statement from fifty-one internationally

recognized scientists in the field of what we know and do not know about intervening in human aging.[28]

Basically, while the proponents of anti-ageing medicine argue for an interventionist, therapeutic approach to ageing, ageing being conceived in this case as a discrete, natural process, biogerontologists suggest that there is nothing either discrete or natural about ageing. The plasticity of the human organism, and in particular how first death and then ageing have been postponed dramatically during the last few centuries, is where biogerontologists find support to not only advance a comprehensive programme of research into the pathways shared by the diverse debilities associated with old age, but also to propose a wide-ranging set of interventions in public health.

When viewed from this perspective, biogerontology is likely to prove very attractive to all those clinical practitioners involved in managing degenerative diseases, from the primary care practitioners controlling their patients' blood pressure to the specialized clinicians required to train these practitioners in the assessment of, for example, the earliest symptoms of Alzheimer's Disease. Furthermore, biogerontology would also appear to offer opportunities of development to a great variety of actors in the wider market for healthcare. There can be little doubt that investigation of the mechanisms involved in the onset of these degenerative diseases greatly expands the potential market for those pharmaceutical companies that have invested in treatments for the ailments of old age, such as the Geron Corporation with respect to Alzheimer's Disease, because the threshold of treatment moves ever backward to encompass a greater fraction of the population. This said, the investigation of these same mechanisms also offers opportunities to those providing the wherewithal and support to secure 'healthy lifestyles' from birth to death. One of the key changes in the organization of research, clinical practice and policy at the end of the twentieth century has been the shift from the 'problem of disease' to the 'problem of health'. The focus is not on restoring health, but on maintaining health and preventing disease into advanced age. Such insistence on the preservation of a good life in old age entails not only constructing an understanding of the molecular, individual and social dynamics that lead to illness, but also a reliance on preventative therapeutic strategies and health promotion programmes. These in turn are sustained by enhanced epidemiological surveillance technologies that regulate access to therapies and programmes by identifying risk factors or states and supporting individuals' reorganization of their conduct in light of such risks. In so doing, they call upon individuals identified through screening programmes and characterized through a variety of molecular and demographic markers to produce and maintain their own health. Viewed in this broadest of all biopolitical contexts, biogerontology can also be regarded as stabilizing a central expectation of contemporary public and private health providers, namely the reduction of degenerative diseases and their associated costs. From these providers' perspective, death as a result of a sudden, massive heart attack or some such 'catastrophic illness' is much preferable to the development of long-term, debilitating and ultimately fatal conditions such as Alzheimer's Disease. All this, in sum, is the deeper implication of the proposed 'new model of health promotion and disease prevention for the 21st century' and its ambitious expectations: 'the exploration of the mechanisms by which ageing can be postponed in laboratory models will yield new models of preventive medicine and health maintenance for people throughout life, and the same research will also inform a deeper understanding of how established interventions, such as exercise and healthy nutrition, contribute to lifelong wellbeing'.[29]

As Luc Boltanski and Laurent Thévenot (2006) might observe, however, this extraordinarily ambitious vision of health care and its future also comes at a cost. By the time of its announcement, de Grey was no longer part of the biogerontological alliance, having moved into the anti-ageing camp. Prompted by de Grey's movement and the opening of the alternative visions for the future of biogerontology that are conveyed by new journals such as *Rejuvenation Research*, Boltanski and Thévenot would argue that the domain of gerontology has so expanded that it now resists any precise delimitation and definition. Contrary to expectation, such openness and the many links that it enables is no guarantee of success because the very process of opening is as much the source of uncertainty and instability as it is the source of innovation. The multiplication of potentially productive connections also calls ever more insistently for clarification about who is and who is not to be included in the expanding vision of the future. In other words, the expansion of the biogerontological commons creates its own divisions.[30]

Governing the contemporary polity

Whether and how the link between the laboratory, preventative medicine and health maintenance programmes proposed by contemporary biogerontology will work in practice is a matter for the future to determine. In the meantime, it is worth noting that, thus far, the proposals to forge such a link have not only proved controversial, but have also been greeted within the administrative structures of the modern state with some reluctance. The response of the British government to the recommendations advanced in a report by the House of Lords' report on *Ageing: Scientific Aspects* (2005) points to some of the issues involved.

In 2004, answering growing political concern about an ageing population, as well as much-publicized expectations that the genomic sciences would transform how such problems might be managed, the House of Lords requested that its Science and Technology Committee appoint a select committee to examine the 'scientific aspects of ageing', and the terms of reference indicated that the Lords were especially interested in the biological processes of ageing and in any promising areas of research that might benefit older people by delaying the onset of long-term illnesses and disabilities. Significantly, however, the executive summary of the final report suggests that the committee was much aggrieved that the Secretary of State for Work and Pensions, despite their official remit to act as 'Champion of Older People', decided not to submit to the select committee any written evidence on the subject.[31] It is also evident that the Lords were very surprised that it was not the science secretary who responded to the publication of report on the 'scientific aspects' of ageing, but the minister for work and pensions. This minister observed simply that 'old age' had long been a major concern of the government and that it had already invested very heavily in the improvement of health and social care, as well as the reform of pensions. The undoubtedly dismissive tenor of the government's response was encapsulated by its concluding, general comments on the report:

> In conclusion, the Government thinks that the Committee has raised important issues and agrees that there are significant challenges to harness the opportunities that science

offers to improve the well-being of older people. However, the Government is confident that it is addressing those issues and challenges. The Government will be happy to continue liaising with the Committee on developments. In addition, the Government has made the Committee's report available to the Devolved Administrations who will have regard to the helpful evidence and conclusions when developing policies related to ageing.[32]

Significantly, citing a memorandum written by Thomas Kirkwood, as one of the principal scientific advisors to the select committee, the Lords wrote in response that:

> It is particularly disappointing that the Government seem to wish to 'pigeon-hole' ageing research, as if ageing were an isolated, discrete problem, and that research into ageing must necessarily compete with research into other areas. Thus the response reproduces the familiar mantra that 'given finite resources, there will always be a need to balance competing priorities for research'. As we sought to demonstrate in our Report – a point repeated by Professor Kirkwood in his written comments – ageing is a continuum, affecting all of us all the time. He also reiterates the point made in our Report, that generic research into the process of ageing, far from being in competition with research into specific conditions affecting older people, may be 'the most direct route to developing novel interventions and therapies'. There is no sign of such holistic thinking in the Government response.[33]

The reference to the adverse impact on the development of 'new interventions and therapies' is noteworthy in light of the government's contemporaneous investment in securing British leadership in the development of a global market for new drugs and new biomedical regimes. More importantly, however, the Lords' contrast between 'specific conditions affecting older people' and the notion that 'ageing is a continuum, affecting all of us all the time' was informed by Kirkwood's observation that: '[T]here are scientific connections between birth, early years, childhood and adolescence that have major impacts on health and quality of life in middle and old age. These need much greater attention.'[34] These views call into question not just the way in which specific fields of medical specialization are organized, but also the organization of the research that should subtend these fields, and do so to such an extent that biogerontological discourse is best understood as offering a veritable critique of contemporary biomedicine. At the same time, however, this driving ambition to secure the integration of the modes of representing and intervening that are derived from the clinic, the laboratory and the practices of preventative medicine may also explain why the Department for Work and Pensions rejected the proposals of the Science and Technology Committee. As the government wrote in its more detailed comments upon the report:

> The Government agrees that there is a need to improve the level of co-ordination between Research Councils and between all funders of research into this area.
>
> The issue of policy leadership on ageing is being addressed at the most strategic levels of Government. Late in 2004, the Government Chief Scientific Adviser's Committee (CSAC) consulted widely on 'grand challenges' facing public policy where scientific research can play a major role in establishing the way forward. Three themes (including ageing) were agreed in March 2005 and Working Groups identified to develop those themes further. This is work in progress. With the support of other Chief Scientific Advisers, working across Departmental boundaries, it was agreed that the natural policy lead fell to the Department of Work and Pensions (DWP). DWP has the Ministerial Champion on Ageing and shares a major stake in the outcome with the Department of Health and others.[35]

The Department for Work and Pensions is dedicated to addressing the issues confronting a specific subset of the population, namely that cohort which is capable of working or once were so capable, currently listing its central activities as 'employment, pensions and ageing society, and welfare'.[36] It is not invested in governing the totality of the life course, from 'cradle to grave', which is what the discovery of 'scientific connections between birth, early years, childhood and adolescence that have major impacts on health and quality of life in middle and old age' entails. To do so would also entail overstepping its powers, or, more likely, allowing others to overstep their own boundaries and proper place. The translation of this construction of the governmental terrain and its associated forms of knowledge is, of course, to endorse an understanding of ageing as a pathological state and the current privileging of Alzheimer's Disease, dementia and the neurosciences as the critical site of intervention.

Rethinking biopolitical governance, rethinking the fate of the mortal organism

How one is to understand the relationship between ageing, the mortal body and its constituent parts in a positive and affirmative manner is today a quite complicated business.

August Weismann once wrote, in *Essays Upon Heredity and Kindred Biological Problems*, that:

> Probably at first the somatic cells were not more numerous than the reproductive cells, and while this was the case, the phenomenon of death was inconspicuous, for that which dies was very small. But as the somatic cells relatively increased, the body became of more importance as compared with the reproductive cells, until death seems to affect the whole individual, as in the higher animals, from which our ideas upon the subject are derived. In reality, however, only one part succumbs to natural death, but it is a part which in size far surpasses that which remains and is immortal, the reproductive cells.[37]

It is tempting to understand this statement as implying that the mortal organism is of secondary importance to understanding the evolution of life. Though the most famous, Richard Dawkins is only one among many proponents of such a reading. When attention is directed towards 'the evolved capacity of somatic cells to carry out effective maintenance and repair', as Thomas Kirkwood and Steven Austad once put it, the ageing organism assumes however unexpected importance.[38] What is then required is the integration of the understanding of ageing developed in both the laboratory and the clinic, as well as in the course of articulating the practices of preventative medicine and health maintenance, aiming to thus secure 'lifelong wellbeing'.[39] There is, in other words, no displacement of the organism by molecules and populations as foundational units of embodied existence, but an attempt instead to articulate three different levels of biological analysis, namely molecules, organisms and populations, as well as their corresponding institutional formations. As this chapter has sought to convey, however, on those occasions when institutions of government have been prepared to listen and to entrust biogerontologists with the exploration of how best to secure the health and longevity of the population, they have not found the resultant proposals wholly persuasive. They are puzzled by interventions which, to them, do not seem fundamentally different from existing modes of governing health, and when they decide to leave matters as they are, biogerontologists' response is that these officials have failed to understand how 'ageing is a continuum, affecting all of us all the time', how 'there are scientific connections between birth, early years, childhood and adolescence that have major impacts on health and quality of life in middle and old age', and how 'there is no sign of [the requisite] holistic thinking'.[40] Such rejection of what would seem like commonsensical notions may be due to the continuing commitment of departments of government to the needs of specific subsets of the population, here, the chronologically aged and their distinctive problems. As Judith Treas (2009) has observed, segmentation of the population into age cohorts may well be the fruit of arbitrary conventions, but such conventions and all their related disciplinary technologies are required to govern the modern state. If this surmise is correct, it indicates that there are significant obstacles to the biogerontological programme and that they stem from the uneasy coexistence of different modes of governance, namely biopolitical and disciplinary, if not molecular as well.

Finally, drawing this second section of *Biopolitics and the Philosophy of Death* to a close, the challenge confronting biogerontology seems very similar to that which contemporary neo-liberal modes of governance seek to address, when, under the rubric of 'joined-up government', they attempt to balance the flexibility and responsiveness that comes of devolving power upon markets and other non-governmental institutions against the demands of strategic direction. If the hierarchical ordering of powers from the centre outwards is supposed to give way to the coordination of these autonomous agents, this requires thinking across what are effectively different and often incommensurable ways of constituting objects of intervention and working with such objects. Just as with biogerontology, this is not proving very easy.[41] Strikingly, in recent reflections on contemporary governmental forms, Mitchell Dean (2013) observes that it is no longer clear what power is, partly as a result of the convergence of biopolitical and neo-liberal modes of governance, and partly as a result of critical practices that seem averse to the notion that power might explain anything. Drawing on a critical reading of Giorgio Agamben's *The Kingdom and the Glory* (2011), a genealogy of the present convergence of sovereignty and economy, Dean suggests that this situation calls for systematic reflection on the very foundations of contemporary assumptions about power, faithful to the 'mystery of power', but resistive to the allure of either transcendent or immanent perspectives on such mystery.[42] In a similar fashion, the next task for *Biopolitics and the Philosophy of Death* is to revisit the foundations of critical reflection upon the forms of life posited by biopolitical modes of thought, particularly in relation to the fate of the mortal organism, paying due attention to both the analysis of the biogerontological project offered thus far and the commonalities between the two enterprises, biological and critical, commonalities that can be traced back to Weismann and Weismannism. One is even tempted to observe that, when viewed from this perspective, the fate of power and the fate of the mortal organism are perhaps inseparable.

CHAPTER 6

Life, death and philosophy

If important obstacles stand in the way of the biogerontological proposal of a new approach to promoting the health and longevity of the modern polity, the argument advanced in earlier chapters of *Biopolitics and the Philosophy of Death* has been that these obstacles stem from the uneasy coexistence of different modes of governance. Overcoming such obstacles requires thinking across what are effectively different ways of coordinating relationships between organisms, molecules and populations. As the previous chapters have also sought to intimate, this is not easy, because ageing and death are not like any other biomedical processes and objects, testing the limits of biomedical knowledge and the corresponding institutional formations. This third part of the argument seeks to draw out the extent to which philosophy might be regarded as prey to the very same difficulties that are involved in offering an affirmative biomedical representation of ageing and death.

While the differences between them are many, biology and modern philosophy are intimately related. In a much-quoted passage of *The History of Sexuality* (1974), Michel Foucault wrote that 'for millennia, man remained what he was for Aristotle: a living animal with the additional capacity for a political existence; modern man is an animal whose politics places his existence as a living being in question'.[1] How this claim is to be understood is a matter of much debate, but it is generally agreed that the transformation of humans' relationship to their embodied existence which took place during the nineteenth century, and that was discussed in previous chapters, as a context to better understand the development of biology and medicine more generally, has proven fundamentally important to the evolution of modern philosophy. It is difficult to underestimate the importance of the biological achievements of the nineteenth century, from Xavier Bichat to Charles Darwin and beyond, to Friedrich Nietzsche, Henri Bergson or Martin Heidegger's philosophical thought.[2] Foucault and Gilles Deleuze are greatly indebted to such thought and all that it owes biology. Strikingly, Foucault brought *The Order of Things* (1966) to a close by foreshadowing the death of Man, writing:

> As the archaeology of our thought easily shows, man is an invention of recent date. And one perhaps nearing its end. If those arrangements [of knowledge sustaining this invention] were to disappear as they appeared, in some event of which we can at the moment do no more than sense the possibility – without knowing either what its form will be or what it promises – were to cause them to crumble, as the ground of Classical thought did, at the end of the eighteenth century, then one can certainly wager that man would be erased, like a face drawn in sand at the edge of the sea.[3]

Drawing attention to this pivotal image, Deleuze argued that Foucault was unable to explain what exactly was destined to disappear and what would take its place because his understanding of embodied existence and the passing historical formation to which he referred his readers could not be disentangled from one another. To clarify the point, Deleuze focused particularly on Foucault's debt to Bichat's distinctive understanding of death, 'the first act of a modern conception of death'.[4] As will be discussed over the course of the next two chapters, Deleuze's criticism and his own response to the challenge he thus set, a response that is inseparable from both August Weismann's ambiguous legacy and the many questions such legacy raises about of the relationship between molecules, populations and the mortality of the organism, are useful to the exploration of the manner in which the contemporary reconfigurations of the relationship between life and death are played out within philosophical discourse, testing its own limits. To this end and more specifically, the present chapter sets out to articulate the differences between Foucault and Deleuze by examining first the relationship between Foucault and Heidegger, focusing particularly on their presuppositions about the biology of death. The chapter will then turn to Deleuze, positioning his views on the excessive importance that Heidegger and Foucault would seem to have attached to human mortality between Bergson and Weismann's understanding of life, death and embodied existence. In so doing, the chapter will draw out how Deleuze's pivotal notion of a 'body without organs' raises issues that are not dissimilar to those confronting the attempt to offer a positive representation of the human organism and its mortality within the domain of the contemporary biomedical enterprise. The full extent and significance of the parallel for contemporary thought is, however, a matter for the next and final chapter of *Biopolitics and the Philosophy of Death*.

Foucault, Deleuze and philosophy

Whether one agrees with their analyses or not, there can be little doubt that Michel Foucault and Gilles Deleuze count among the most influential philosophers of the later twentieth century. Their shared concern about the nature of embodied existence and its importance to any understanding of modern thought is inseparable from their equally shared, trenchant and far-reaching critique of the modern age as an unparalleled stage in the history of political domination. Today, they maintain, endlessly proliferating discourses and institutions infiltrate all aspects of everyday life, everywhere determining norms and the bounds of normality. The attention thus paid to the play of power and knowledge certainly explains the interest in Foucault and Deleuze's work not just within, but also outside philosophy. Crucially, Foucault and Deleuze also share the view that the elimination of the subject posited by modern thought, that form of thought that is founded on the assumption of a singular and autonomous centre of reflection, is the only adequate response to such pervasive domination.[5]

At the same time, Foucault and Deleuze's perspectives on the matters and propositions just set out are also very different. Briefly, where Foucault emphasized both the manner in which regimes of power and knowledge target the body and how the deployment of such power rests on modern disciplinary technologies, Deleuze was much more interested in the shaping of desire. While desire was, for Foucault, a product of modern disciplinary technologies, Deleuze regarded desire as fundamentally important to understanding the capacity of bodies to proliferate and exceed all boundaries, even the boundary between life and death. In other words, where Foucault tended toward a comprehensive critique of modernity, Deleuze sought to characterize and appropriate the liberating aspects of the modern order, particularly its extraordinarily disruptive capacities. As Karl Marx, who in many ways was Deleuze's touchstone, put it in his powerful paean to the transformative power of capital:

> Constant revolutionising of production, uninterrupted disturbance of all social conditions, everlasting uncertainty and agitation distinguish the bourgeois epoch from all earlier ones. All fixed, fast-frozen relations, with their train of ancient and venerable prejudices and opinions, are swept away, all new-formed ones become antiquated before they can ossify. All that is solid melts into air, all that is holy is profaned, and man is at last compelled to face with sober senses his real conditions of life, and his relations with his kind.[6]

In a similar fashion, if Aubrey de Grey can announce quite assuredly that 'the first person to live to be a thousand years old is certainly alive today', it is because, as seen in previous chapters, modern biology, biomedical technologies and biopolitical governance are transforming the relationship between life and death. In this context, philosophers inspired by Deleuze, such as Rosi Braidotti, for example, can argue equally credibly that any insistence on the fundamental importance of mortality as the starting point for any philosophical reflection upon the human condition is wholly misguided.[7] As observed earlier, Braidotti's argument is directed specifically at Heidegger and all those philosophers indebted to Heidegger's centrally important understanding of existential awareness as predicated on human finitude, to Heidegger's 'theory of Being as deriving its force from the annihilation of animal life'.[8] Arguably, Foucault must be counted among these philosophers, even if the exact relationship between Foucault and Heidegger has been much discussed ever since the publication of Hubert Dreyfus and Paul Rabinow's *Foucault, Beyond Structuralism and Hermeneutics* (1983).[9]

Heidegger, Foucault and the biology of death

To put the matter in the simplest terms possible, Michel Foucault's relationship to Martin Heidegger's understanding of embodied existence is best characterized as thoroughly ambiguous.

Participating in a more general transformation of cultural sensibilities toward the boundaries between human life and death, Heidegger sought to advance a comprehensive reconstruction of philosophy by positing a reflective form of existence that was not only embodied, but was also produced by the inescapable immanence of death within life itself, *Dasein*.[10] *Dasein* emerged in the cultivation of the understanding of its own death as impending possibility and as an understanding that no one else could possibly share. 'Death', as Heidegger put it, 'reveals itself as that possibility which is one's own-most, which is non-relational, and which is not to be outstripped.'[11] This, according to Heidegger, was the necessary foundation of authentic existence and the ground upon which to reconstruct the entirety of philosophy. Importantly, Heidegger, like Friedrich Nietzsche before him, sought to purge philosophical argument of all metaphysical assumptions by grounding his understanding of human existence within contemporary biology, so much so that this understanding of death as internal to life was indebted explicitly to Eugen Korschelt's *The Duration of Lifespan, Ageing and Death* (1922).[12] In this book, Korschelt, one of August Weismann's students and target of Alex Comfort's most uncompromising criticism, reported upon his investigations of the diverse mechanisms that enabled multicellular organisms to live despite the constant turnover of the component cells, a turnover that could not be regarded as a form of mortality because Korschelt, like many of his contemporaries, regarded death as only applying to complex forms of life. Death, in other words, was only relevant to the organism and marked the life of the organism as different to the course of life more generally.[13] In some ways, this is the very same understanding that Edmund Cowdry sought to secure between the 1930s and 1950s. This said, it is equally important to recall that Heidegger was deeply concerned about the relationship between philosophy and biology, which he thus advanced, because it threatened to subordinate the former to the latter. As David

Farrell Krell (1992) has argued, one must understand the evolution of Heidegger's thinking about the 'anxious animal' as the product of complex negotiations. These negotiations sought to ground existential awareness in the mortality of the body, while also seeking at the same time to avoid any reduction of the issue to a matter of biology and so avoid any repetition of Nietzsche's arguably mistaken collapse of all difference between philosophy and biology.[14] The difficulties involved were none more visible than when Heidegger wrote in *Being and Time*:

> Let the term 'dying' stand for the way of Being in which Dasein is towards death ... [W]e must say that Dasein never perishes. Dasein, however, can demise only as long as it is dying. Medical and biological investigation into 'demising' can obtain results which may even become significant ontologically if the basic orientation for an existential Interpretation of death has been made secure. Or must sickness and death in general – even from a medical point of view – be primarily conceived as existential phenomena?[15]

Distinguishing between perishing, demising and dying, let alone disentangling these terms from the experience of illness, is not easy, to say the least, and the difficulty has been a topic of much discussion about Heidegger's arguably lamentable anthropocentrism.[16] More importantly, however, Heidegger's negotiation was facilitated, at least in part, by his reliance on Korschelt because, like Weismann, 'his old master', Korschelt understood the death of the organism to play a structural and beneficial role, namely to advance 'the further development and propagation of life'.[17] Such understanding enabled Heidegger to attribute a positive and productive meaning to death, beyond negation, and to do so in a manner wholly consistent with, but not wholly reducible to some understanding emerging from within the contemporaneous development of biological thought.

As was observed in the introduction, Foucault viewed Heidegger's reconstruction of philosophy as tacitly reintroducing the very metaphysical assumptions it sought to overcome. Since, as Heidegger insisted, the pivotal notion of one's own death could not be rendered as a positive object of human inquiry, but was to be referred instead to some phenomenological structure, Foucault could not but conclude that Heidegger's entire project was predicated on the tacit presupposition of an experiencing subject, the very entity whose emergence they both sought to understand as the product of historical processes.[18] This said, Foucault's own understanding of death and mortality suggests that the two philosophers' perspectives on the relationship between life, death and the production of the subject could also be regarded as quite close.[19]

Over a century ago, Nietzsche famously announced the death of God, seeking to thus capture the abandonment within modernity of all belief in the possibility of a superhuman, transcendental perspective. As Gilles Deleuze observes in his critical review of Foucauldian thought, Foucault pressed the Nietzschean argument further, by transforming reason itself into an object of historical inquiry. In the course of doing so, Foucault articulated how the modern subject was fated to discover the fundamental impossibility of understanding itself, so undoing all confidence in the Cartesian *dictum* 'I think, therefore I am'.[20] While Deleuze acknowledged the importance of the argument thus advanced by suggesting that the culmination of the historical process which Foucault had outlined in *The Order of Things* (1966) amounted to nothing less than the death of Man, he also suggested that Foucault had been unable to delineate the contours of the new site of thought and deliberation that would take the place of Man because he remained wedded to the understanding of embodiment that he had first adumbrated in *The Birth of the Clinic* (1964).[21]

In *The Birth of the Clinic*, Foucault argued that the emergence of a distinctively modern understanding of life is inseparable from the fundamental reconfiguration of the relationship between life and death, which he associated with Bichat's famous definition of life as 'the sum of all activities that resist death'.[22] Foucault also observed how the articulation of these physiological activities rested on Bichat's deployment of death as 'the great analyst that shows the connexions by unfolding them, and bursts open the wonders of genesis in the rigour of decomposition'.[23] Over the ten years between the writing of *The Birth of the Clinic* and *The History of Sexuality*, this power of death to disclose the truth of human, embodied existence was transformed into the modern triangle of life, knowledge and power. Death, in this configuration, demarcated the subject's interiority and freedom. As Foucault put it, while characterizing the emergence of modern biopolitical governmentality:

> One might say that the ancient right to take life or let live was replaced by a power to foster life or disallow it to the point of death. This is perhaps what explains that disqualification of death which marks the recent wane of the rituals that accompanied it. That death is so carefully evaded is linked less to a new anxiety which makes death unbearable for our societies than to the fact that the procedures of power have not ceased to turn away from death. In the passage from this world to the other, death was the manner in which a terrestrial sovereignty was relieved by another, singularly more powerful sovereignty; the pageantry that surrounded it was in the category of political ceremony. Now it is over life, throughout its unfolding, that power establishes its dominion; death is power's limit, the moment that escapes it; death becomes the most secret aspect of existence, the most 'private'. It is not surprising that suicide, once a crime, since it was a way to usurp the power of death which the sovereign alone, whether the one here below or the Lord above, had the right to exercise became, in the course of the nineteenth century, one of the first conducts to enter into the sphere of sociological analysis; it testified to the individual and private right to die, at the borders and in the interstices of power that was exercised over life. This determination to die, strange and yet so persistent and constant in its manifestations, and consequently so difficult to explain as being due to particular circumstances or individual accidents, was one of the first astonishments of a society in which political power had assigned itself the task of administering life.[24]

While the contemporary calls to secure the right to die, to which these words point, are a matter for the next chapter, it is then difficult to turn a deaf ear to Heidegger's voice in Foucault's famous observation, in *The Order of Things*, that:

> [M]an has not been able to describe himself as a configuration in an episteme without thought at the same time discovering, both in itself and outside itself, at its borders yet also in its very warp and woof, an element of darkness, an apparently inert density in which it is embedded.[25]

Admittedly, there is some difference between Bichat's and Weismann's understanding of mortality, but it is important to recall the ambiguities of the latter's understanding of the organism and its significance. Equally, there is some difference between Foucault's detached observations about the constitution of the modern subject's psychic structure and Heidegger's ontological preoccupations, but, as many critics have observed, the difficulties involved in maintaining such a detached position from one's own historical situation are considerable, especially where epistemic formations are supposed to be all-encompassing.[26] Deleuze, recognizing the proximity between Heidegger and Foucault, tries to disentangle the two by seeking to transform Foucault into a vitalist.[27]

This understanding of Foucault as in the thrall of death, from which he must be liberated, motivates current endeavours to align Foucault and Deleuze. Deleuze would appear to offer a very different perspective on the relationship between life, death and subjectivity, and it is this difference that Rosi Braidotti and others find most attractive. The hope is that the endeavour will result in a more affirmative account of biopolitics than the constitutive role of mortality and finitude will afford.[28]

Deleuze between Weismann and Bergson

Gilles Deleuze has no time for the importance that Martin Heidegger and Michel Foucault attach to death and finitude, blind as it is to the creative play of chance, however violent, in 'the domain of … partial deaths, where things continually emerge and fade (Bichat's zone)'.[29] Like Heidegger before him, Deleuze sought to reconstruct the entirety of philosophy, but in the manner of an unrepentant metaphysician, intent upon establishing a more affirmative account of life. He did so by inverting Baruch Spinoza's rejection of dualism, which posited that everything that exists should be understood as the modification of a single and universal substance, God. Deleuze proposed instead that there was no one substance, but only a process of perpetual differentiation and proliferation. Significantly, August Weismann's germ and the diversity of forms that the germ generates in the process of variation provided a resonant way of thinking about this process. Thus, in *Anti-Oedipus* (1972), Deleuze and Félix Guattari, Deleuze's sometime collaborator, sought to advance an alternative analysis of contemporary socioeconomic order and its historical constitution by recasting the understanding of desire. Desire, according to Deleuze and Guattari, was to be understood not as a response to something lacking, but as an autonomous and creative agent, just like Spinoza's *conatus*, driving the creativity of nature, understood here not as *natura naturata*, but as *natura naturans*.[30] Such understanding entailed an equally different relationship between the unity of the point of origin and the multiplicity and diversity of the present historical moment. To this end, Deleuze and Guattari turned to Dogon cosmogony, whereby the genetic relationship between the origin and the genesis of contemporary Dogon people allows for differentiation and yet maintains a materially immediate bond across the generations and with the point of origin, Amma's 'cosmic egg'. Deleuze and Guattari wrote:

> Yes, I have been my mother and I have been my son. It is rare that one sees myth and science saying the same thing from such a great distance: the Dogon narrative develops a mythical Weismannism, where the germinative plasma forms an immortal and continuous lineage that does not depend on bodies; on the contrary, the bodies of the parents as well as the children depend on it. Whence the distinction between two lines, the one continuous and germinal, but the other discontinuous and somatic, it alone being subjected to the succession of generations.[31]

Deleuze's distinctive inclination toward dismissing any preoccupation with death seems to be best understood as indebted to this Weismannian distinction between the germinal and somatic components of the organism, albeit in a manner that is attentive to the complex relationship between Deleuze, Weismann and Henri Bergson, the latter being one of Deleuze's chief sources of philosophical inspiration.[32]

Writing in the wake of Charles Darwin's *On the Origin of Species* (1859), Bergson, one of the great, if now overlooked, philosophers of the early twentieth century, regarded nature as the domain of creation and the production of novelty. The aim of *Creative Evolution* (1911), perhaps Bergson's most famous book, was to offer a philosophical perspective capable of accounting for both the continuity of life and the discontinuity implied by creation and the production of novelty. More specifically, in his earliest work, Bergson had sought to reject the foundational understanding of all reality as determined by necessity, as well as the accompanying reduction of all transformative processes to static natures or essences. As a result, Bergson was greatly attracted to Darwin's evolutionary account of organic diversity. Darwin's theory captured the irreducible flux of 'becoming'. At the same time, Bergson was forced to reject Darwin's mechanism of natural selection because, like Friedrich Nietzsche, he observed how it reintroduced an essentially static and negative determining agency, selecting some forms and eliminating others. From Bergson's perspective, everything that might be said to exist was in a constant state of flux, subject to a universal, immanently determined process of perpetual transformation. Everything was duration, always caught up in the passage of time, and such duration was nothing else but the movement of freedom. It was on this basis that Bergson sought to go beyond deterministic, Darwinian conceptions of evolution to the creative conception that lends *Creative Evolution* its title. Unsurprisingly, Bergson was then driven to criticize Weismann's response to the problems plaguing Darwin's theory because Weismann's notion of an unchanging germ seemed to treat the development of novel forms as determined by preceding ones, so that there was no real change or creation. As Bergson put it in *Creative Evolution*:

> It is well known that, on the theory of the 'continuity of the germ-plasm', maintained by Weismann, the sexual elements of the generating organism pass on their properties directly to the sexual elements of the organism engendered. In this extreme form, the theory has seemed debatable, for it is only in exceptional cases that there are any signs of sexual glands at the time of segmentation of the fertilized egg. But, though the cells that engender the sexual elements do not generally appear at the beginning of the embryonic life, it is none the less true that they are always formed out of those tissues of the embryo which have not undergone any particular functional differentiation, and whose cells are made of unmodified protoplasm. In other words, the genetic power of the fertilized ovum weakens, the more it is spread over the growing mass of the tissues of the

embryo; but, while it is being thus diluted, it is concentrating anew something of itself on a certain special point, to wit, the cells from which the ova or spermatozoa will develop. It might therefore be said that, though the germ-plasm is not continuous, there is at least continuity of genetic energy, this energy being expended only at certain instants, for just enough time to give the requisite impulsion to the embryonic life, and being recouped as soon as possible in new sexual elements, in which, again, it bides its time. Regarded from this point of view, life is like a current passing from germ to germ through the medium of a developed organism. It is as if the organism itself were only an excrescence, a bud caused to sprout by the former germ endeavouring to continue itself in a new germ. The essential thing is the continuous progress indefinitely pursued, an invisible progress, on which each visible organism rides during the short interval of time given it to live.[33]

In other words, Bergson maintained that, while there is continuity of life, there is no material essence transmitted unchanged from one generation to the next because the transforming organism provides the material transmitted. This said, Bergson was just as critical of the alternative proposed in contemporaneous orthogenetic accounts of life because they too entailed that future forms were determined beforehand. Bergson sought to develop an intermediate form of teleology. Expanding upon the above understanding of continuity, his answer was the notion of an *élan vital* driving the diversification and complexification of all living organisms. This was the only way, Bergson argued in his much-overlooked *Two Sources of Morality and Religion* (1932), that humanity could hope to overcome a deterministic and ultimately metaphysical understanding of the natural world, or, in other words, 'by slightly distorting the terms of Spinoza', this was the only way to 'get back to *natura naturans*'.[34]

Significantly, *Two Sources of Morality and Religion* also offers important insight into Bergson's understanding of human mortality and helps to illustrate how Deleuze comes to construct his distinctive philosophical edifice by combine Bergson's and Weismann's different understanding of evolutionary process. Advancing a distinction similar to Heidegger's between perishing, demising and dying, Bergson argued that:

Animals do not know that they must die. Doubtless some of them make the distinction between the living and the dead; we mean by this that the sight of a dead creature and of a living one does not produce in them the same reactions, the same movements, the same attitudes; this does not imply that they have a general idea of death, any more than they have of life, or any general idea whatsoever, at least in the sense of a mental picture and not simply a movement of the body.[35]

Knowledge of mortality would anyway be 'useless'.[36] The situation was entirely different when it came to the human animal, however. As Bergson put it, echoing Heidegger's notion of the 'anxious animal', but in a manner that was underpinned by a wholly different logic concerning the human animal's intelligence and inventiveness:

> [M]an knows he will die ... Seeing that every living thing about him ends by dying, he is convinced that he will die too. Nature, in endowing him with intelligence, must inevitably lead him to this conclusion. But this conviction cuts athwart the forward movement of nature. If the impetus of life turns all other living creatures away from the image of death, so the thought of death must slow down in man the movement of life ... [I]t is at first a depressing thought, and would be more depressing still, if man, while certain that he must die, were not ignorant of the date of his death. Death is indeed bound to come, but as we are constantly becoming aware that it does not come, the continued repetition of the negative experience condenses into a barely conscious doubt, which diminishes the effect of the reasoned certainty. It is none the less true that the certainty of death, arising at the same time as reflexion in a world of living creatures constructed to think only of living, runs counter to nature's intention.[37]

If this anxious state was an impediment to continued progress of human evolution, Bergson regarded the invention of religion as an answer to the problem:

> To the idea of inevitable death she [nature] opposes the image of a continuation of life after death; this image, flung by her

into the field of intelligence, where the idea of death has just become installed, straightens everything out again … Looked at from this … standpoint, religion is a defensive reaction of nature against the representation, by intelligence, of the inevitability of death.[38]

In other words, God was an invention designed to ensure the continued proliferation of biological life and creativity, which one day would even defeat death itself. As Bergson put it in *Creative Evolution*:

> As the smallest grain of dust is bound up with our entire solar system, drawn along with it in that undivided movement of descent which is materiality itself, so all organized beings, from the humblest to the highest, from the first origins of life to the time in which we are, and in all places as in all times, do but evidence a single impulsion, the inverse of the movement of matter, and in itself indivisible. All the living hold together, and all yield to the same tremendous push. The animal takes its stand on the plant, man bestrides animality, and the whole of humanity, in space and in time, is one immense army galloping beside and before and behind each of us in an overwhelming charge able to beat down every resistance and clear the most formidable obstacles, perhaps even death.[39]

Paradoxically, Weismann, Bergson's nemesis, maintained that death was a product of evolution and answer to a problem, namely how best to ensure the continued proliferation of life, and not the fundamental and ubiquitous biological process which Bergson seemed to assume. Just as Bergson rejected the Darwinian notion of natural selection because, from his point of view, it seemed to play a wholly negative role in shaping the diversity of life, Deleuze turned to Weismann to press further Bergson's argument. As the next section will evince, Weismann's distinction between the germinal line and the somatic body served as a powerful model for the development of Deleuze's philosophical project, but Deleuze's turn to Weismann was always shaped by his greater debt to Bergson and Bergson's belief in the unlimited, creative powers of life.[40]

The mortal organism and the body without organs

Over the course of writing *Bergsonism* (1966), *Difference and Repetition* (1968) and *The Logic of Sense* (1969), works central to the Deleuzian œuvre, Gilles Deleuze developed the critically important notion of a 'body without organs', which served increasingly to ground materially Deleuze's critique of contemporary philosophy and his renewal of philosophical vitalism.

From the earliest stages of his philosophical education and just like Michel Foucault, Deleuze was interested in the genesis and transformation of conceptual frameworks. He thus sought to overcome the common presupposition of a transcendental field conditioning empirical experience and a central focus of experience. On this understanding, such focus of experience was understood as the site of divergence and differentiation. As a result, difference was regarded as secondary attribute of substance, rather than the fundamental feature of all that exists. In *Bergsonism*, Deleuze seeks to articulate an alternative understanding of the transcendental field such that it might become the original site of divergence, differentiation and transformation. He does so by first examining how the transcendental field comes to be understood as the domain of prior possibilities. He observes how one tends to think of the real, then of some modification of the real's existence, projecting such modification into a conceptually anterior space and so constituting a set of prior and determinate possibilities. As a result, the real is not only the poorer relative of the transcendental field, which, on this conventional understanding, is constituted as the realm of ideal forms, but the emergence of anything truly novel is also rendered wholly unintelligible because it already exists, as a determinate possibility yet unrealized. Deleuze seeks to overturn this edifice, in its entirety. He does so by turning to Bergson's notion of the virtual. Like the ideal, the virtual conditions experience, but the virtual does not enjoy any determinate identity comparable to the ideal. Determinate identity is instead the product of processes whereby an ever-differentiating field takes material form. In other words, this alternative field is not to be understood as the prior condition required for experience, but as emerging in the very genesis of experience, experience being the movement of actualization, rather than the instant of a prior possibility being realized. As such, the difference between the virtual and the actual is not reality, but material form. Deleuze also adopts from Bergson the notion of multiplicity to characterize the purely differential field he thus constructs, a use whereby the term is intended to designate substance, rather than a numerical qualification of substance.[41]

In *Difference and Repetition*, Deleuze further advanced the argument about the differential field. Importantly, drawing in part upon a reading of August Weismann's *Essays Upon Heredity and Kindred Biological Problems* (1891), Deleuze begins to give the argument greater material consistency.[42] He proposes that the individuation of concretely existing entities involves three domains, namely the virtual, the intensive and the actual, such that intensive processes similar to those encountered in biology are involved in the actualization of a multiplicity into localized and individuated substance. Importantly, if this approach to lending some material form to the play of multiplicity is immune to any charge of idealism, it is because Deleuze, conscious of the need to purge Weismannism of its idealism, draws on Albert Dalcq and Raymond Ruyer's developmental biology, so coming to regard each organism that is formed in the unfolding of these intensive processes as determined by the totality of both evolutionary history and ecological context. The consequent relationship between the organism and the emerging notion of a 'body without organs' is not that of endless repetition of an original genetic plan, as is posited by the neo-Darwinian reading of Weismann, but one where the process of actualization transforms the course of evolutionary history. True to the same neo-Darwinian reading of Weismann, however, the organism is to be regarded as a 'solution to a problem', namely to enable the proliferation of the multiplicity.[43] In this context, the death of any particular organism is to be understood as devoid of any essential significance. As Deleuze puts it:

> As the movement of life shows, difference and repetition tend to become interiorised in signal – sign systems both at once. Biologists are right when, in posing the problem of heredity, they avoid allocating distinct functions, such as variation and reproduction, to these systems, but rather seek to show the underlying unity or reciprocal conditioning of these functions.

At this point, the theories of heredity necessarily open on to a philosophy of nature. It is as if repetition were never the repetition of the 'same' but always of the Different as such, and the object of difference in itself were repetition. At the moment when they are explicated in a system (once and for all) the differential, intensive or individuating factors testify to their persistence in implication, and to the eternal return as the truth of that implication. Mute witnesses or dark precursors which do everything – or at least, it is in these that everything occurs.[44]

The system or organism, in other words, is no more than a site wherein diverse intensive movements intersect momentarily, and the course of the singular life thus actualized is not to be confused with the history of life itself.

If the organism cannot provide any foundation for a reconstruction of philosophy because it is no more than a historical invention, just like Man and God, Deleuze's *The Logic of Sense*, where the body without organs finally appears fully fledged, also suggests that the organism remains nonetheless important. In *The Logic of Sense*, Deleuze draws on Lewis Carroll's *Alice's Adventures in Wonderland* (1865) and *Through the Looking Glass* (1871) to articulate an analysis of the relationship between words, meaning and the body. Carroll's great gift, according to Deleuze, was to be able to disclose the constitutive frailty of the mutual dependence of words, meaning and the body, without falling into madness comparable to that which Antonin Artaud experienced while fathoming the depths of such dependence. Deleuze's argument is, first, that meaning emerges on the boundary between words and things because meaning is expressed in linguistic propositions, but is not to be confused with these same propositions. He also argues, secondly, that the determinate dimensions of such propositions, that is, their capacity for denotation, manifestation and signification, are not just fixed on this boundary, but their articulation depends on a complex relationship to the body. Linguistic expression and the body are inseparable, but the articulation of the determinate dimensions of any meaningful linguistic proposition also rests on separation of the two. Deleuze suggests that Artaud's schizophrenia found its expression in utterances that collapsed the boundary between speech and other bodily sounds, and not just oral sounds.

Significantly, the material substratum upon which this moment when speech is evacuated of all sense is inscribed also offers the first a glimpse of the body without organs. As Deleuze puts it, to this moment 'in which all literal, syllabic, and phonetic values have been replaced by values which are exclusively tonic and not written ... a glorious body corresponds, being a new dimension of the schizophrenic body, an organism without parts which operates entirely by insufflation, respiration, evaporation, and fluid transmission (the superior body or body without organs ...)'.[45] Playing with developmental psychology, Deleuze also maintains that this emergent dimension can be regarded in one of two ways, either as a reversion to an amorphous, pre-embryonic state, or as a fluid state wholly open to transformation. What is at stake, in other words, is a choice between two very different perspectives on the death drive and its role in the constitution of the subject. Where Sigmund Freud posited that the subject is produced at the intersection of two drives, an erotic drive and the drive to return to a primordial, impassive state, Deleuze rejects this understanding of the relationship between life and death. According to Deleuze, the death drive is far more important than Freud appreciates, because, by bringing the life of the individual to an end, death reinvigorates the endless process of differentiation and transformation, that is, it reinvigorates life itself.[46] Thus, Deleuze refers his readers to Joë Bousquet, the poet who, despite being paralysed and in constant pain, rejected the notion of suicide as reifying death and preferred instead to take charge of life by accepting and embracing his pain, which, according to Deleuze, exemplifies openness to death as unfathomable event of transformation. Bousquet once wrote: 'To my inclination for death ... which was a failure of the will, I will substitute a longing for death which would be the apotheosis of the will ... Become the man of your misfortunes, learn to embody their perfection and brilliance.'[47] Reading these words along with Maurice Blanchot's own reflections on death, its singularity and the consequent meaninglessness of suicide, Deleuze adds:

> Nothing more can be said, and no more has ever been said: to become worthy of what happens to us, and thus to will and release the event, to become the offspring of one's own events, and thereby to be reborn, to have one more birth, and to break one's carnal birth – to become the offspring of one's events and

not of one's actions, for the action is itself produced by the offspring of the event.[48]

In other words, Deleuze rejects any consecration of death and seeks instead to transform the subject's self-understanding by evoking the possibility of aligning self and the continuity of life, beyond the mortality of the human body, by evoking the possibility of endless rebirth. This said, it is not clear how this poetic injunction relates to Deleuze's concluding reflections on Émile Zola's *The Beast Within* (1890). Reflecting on the novel's play of biography, culture, and the brutal, but creative forces of nature, Deleuze wrote in an inescapably Weismannian tone:

> Two unequal coexisting cycles interfere with each other: small heredity and grand heredity, a small historical heredity and a great epical heredity; a somatic heredity and a germinal heredity.[49]

For all the vitality and optimism that Deleuze's words exude, it is not clear how Artaud's, Bousquet's and Blanchot's poetic affirmations of life, beyond all reference to death, come together to effect any material transformation within the grander of the two narratives, except by eliding the qualitative differences between the meaning of life at two very different scales and two very different registers, material and symbolic.[50]

In sum, and recalling Titian's 'The Three Ages of Man', the trajectory of the organism from birth to death, as Deleuze seeks to construct it, would seem to be related very ambiguously, if not obscurely, to the course of life at the ecological and evolutionary scales. Equally, the theological overtones betrayed by the old man contemplating the meaning of his mortality, in the form of the *memento mori* in his hand, would seem to find their counterpart in Deleuze's evocation of a 'great epical heredity'.

Weismann and the philosophy of biology

Gilles Deleuze and Félix Guattari announce in *A Thousand Plateaus* (1980) that the 'organism is the enemy'.[51] This seemingly uncompromising statement is wholly consistent with the understanding of the relationship between life, death and the body that Deleuze advances in *Bergsonism*, *Difference and Repetition* and *The Logic of Sense*. At the same time, as was concluded above, the understanding of systems and organisms as 'mute witnesses or dark precursors which do everything – or at least, it is in these that everything occurs' is ambiguous. While such resonant ambiguity will be discussed at much greater length in the next chapter, it also calls for greater attention to both Deleuze and Guattari's discussion of filiation as a trace of the manner in which the actualization of the organism in the mode of repetition remains important to the history of the *phylum*, to the historical trajectory of desire, and to the more general resonance of such conceptual complexity.

As Deleuze and Guattari put it, drawing on contemporary biological notions of coevolution and the viral transmission of genetic information, the proliferation of the body without organs does not depend on any form of filiation:

> Becoming is not an evolution, at least not an evolution by descent and filiation. Becoming produces nothing by filiation; all filiation is imaginary. If evolution includes any veritable becomings, it is in the domain of symbioses that bring into play beings of totally different scales and kingdoms, with no possible filiation. If there is originality in neo-evolutionism, it is attributable in part to phenomena of this kind, in which evolution does not go from something less differentiated to something more differentiated, in which it ceases to be a hereditary filiative evolution, becoming communicative or contagious ... Becoming is a rhizome, not a classificatory or genealogical tree.[52]

Yet, for all their claims to thus dispense with any requirement for mediation by either organism, species or phylum, each linked to the other by hierarchies of repetition, Deleuze and Guattari also seem unable to dispense completely with this reproductive mechanism. They also write that '[p]ropagation by epidemic, by contagion, has nothing to do with filiation by heredity, even if the two themes intermingle and require one another'.[53] The ambiguity of such requirement obtains because the organism, following August Weismann, is a solution to a problem, the propagation of life, which is secured by splitting vital processes into two axes of transformation, the somatic axis associated with the mortal organism, and the germinal axis associated with the continuity of the phylum. As such, the organism is also a functional component of this solution. In other words, if there are problems to be solved, including how to guarantee the continuity of the play of differentiation, it is not clear how they can be disentangled from a nodal centre of transformation and how this is to be separated from the punctuating discontinuity marked by the organism's mortality and finitude. This difficulty presumably lies at the heart of the notion of 'mute witnesses or dark precursors which do everything – or at least, it is in these that everything occurs'. Given the present historical situation, and coming full circle, this difficulty would seem to be precisely the challenge confronting those biologists exercised today by the relationship between genes, organisms and populations, when they ask 'does biology need an organism concept?'[54]

Philosophers of biology such as James Griesemer are greatly concerned about the place of the organism within the explanatory strategies of evolutionary biology. Griesemer has thus observed that:

> Darwin's achievement was to argue successfully that evolution had occurred in nature and that natural selection is an important, perhaps the most important, mechanism of modification operating within generations. Darwin's theory is commonly characterized as an 'umbrella theory', unifying explanations of diverse biological phenomena. However, many biologists have questioned whether the theory is umbrella enough, in the mathematical form given it by population genetics, to cover phenomena from the lowest molecular to the highest taxonomic and ecological levels.[55]

It is not clear, in other words, whether the neo-Darwinian mobilization of August Weismann's work to separate the life of the organism from that of the phylum is capable of encompassing the extraordinary diversity of biological phenomena. More specifically, to the extent that processes of replication can be separated from processes of development such as those that are involved in the transformation of the embryo into the fully formed, reproducing organism, these processes of replication play a central role in a variety of phenomena, from the role of genes in the evolutionary transformation of organisms and biological phyla to the evolution of cultures and social structures. To encompass such variety, neo-Darwinian biologists such as Richard Dawkins have sought to extend the definition of both the replicating unit and the phenotype so that the set of replicating units now includes not only genes, but also structures that are usually regarded as components of an organism's phenotype or an organism's environment. Not only are pivotal distinctions that Weismann, and others after him, introduced to make sense of the diversity of life thus eroded, but such redefinition also divides biologists between those who argue that replication in this extended sense is necessary for the process of selection to operate effectively, and others who argue otherwise. Furthermore, some biologists maintain that selection can occur in the absence of replication, and others still maintain that the diversity of phenomena included in these discussions is best explained without any reference whatsoever to notions such as replication, selection and evolution. Consequently, during the past two decades, a number of biologists and philosophers who work very closely with these biologists have sought to clarify the conceptual foundations of the contemporary biological enterprise. Griesemer himself is interested particularly in the role of replication in the process of natural selection, wishing to free biological thought from the hold of genes and 'molecular Weismannism'. On the understanding that these biologists and philosophers have sought to develop, one could start just as well from the observation that most organisms do not arise already capable of reproduction, but, as Henri Bergson also observed, must undergo an orderly series of transformations before they can reproduce. Inheritance thus ceases to be a property of any component parts, and becomes instead a systemic property. Inheritance also becomes a capacity that is acquired during the course of development. Griesemer has thus come to regard Dawkins's very different understanding of inheritance as based on an analogy with the process of copying, which overlooks how reproducing apparata require an overlap of material systems to be able to produce the copy. From this critical perspective, replication is the endpoint to an extended process of reproduction, only occurring once coding mechanisms are in place, the latter step itself being a form of organic multiplication.[56] Griesemer thus speaks of a process of 'progeneration' to capture the material overlap between parental and filial generations. In so doing, one cannot but wonder whether these philosophers and biologists are not converging upon something like Deleuze's notion of 'perplication', wherein the materialization of abstract ideas results in a proximal and lateral ordering of entities displacing hierarchical alternatives.[57] On the other hand, if all the questions posed by Weismannism and neo-Darwinism about the ordering of biological phenomena are inseparable from the mathematical models and abstract formalisms they employ to connect the dynamics of molecules and populations, what is also at issue in these critiques is modelling itself and its relationship to the objects modelled. As Griesemer puts it, after reviewing the relative merits of genetic explanations of biological transformation and the epigenetic alternatives which he seeks to advance:

> Because Darwinian evolutionary theory is framed in terms of a requirement for a capacity or disposition of heritability it leaves open questions of which inheritance mechanisms are involved in different instances. Because molecular biology is concerned with mechanisms that frequently or generally operate in gene regulation and development (though not without interest in variation), it leaves open questions of which mechanisms are significant in evolution and in what ways. Evolutionary biology is concerned with both the distribution of traits (including the presence of mechanisms) among taxa and the dynamical role of mechanisms in evolutionary processes of change within and among lineages. Thus, in asking questions about reductionism and relative significance of epigenetic phenomena in relation to well-established theories of genetic information we have to ask, 'reduction for whose purpose(s)?' and 'significance relative to whose problems?'[58]

What is at issue, in other words, is not so much a divergence consequent upon different ontological assumptions, as an argument about how best to accommodate the plurality of epistemic practices and explanatory strategies involved in understanding the transformation of biological forms and structures. Thus, if John Protevi (2012) and other neo-Deleuzians look to current discussions within contemporary biology about how best to understand the relationship between genetic mechanisms and developmental processes with great interest, the focus on epistemological and disciplinary structures should urge some caution about too ready alignment of such discussions with Deleuzian metaphysics. The relationship between biology and modern philosophy is fraught with difficulties.

Weismann, biophilosophy and the philosophy of biology

This chapter has sought to outline the importance of biology to the development of Michel Foucault and Gilles Deleuze's philosophical thought, situating their understanding of the relationship between life and death in relation not just to Friedrich Nietzsche, Henri Bergson and Martin Heidegger, but also in relation to August Weismann. There is, of course, an entire philosophical tradition that tends to discount the importance of all these figures, but even within the most rigorous traditions of analytical philosophy, the questions posed by biology have proven very important to the historical development of its distinctive mode of critical inquiry.[59] Thus, for example, Karl Popper, in *The Poverty of Historicism* (1957), dismissed the scientific credentials of evolutionary theory because it did not conform to the requirements of the hypothetico-deductive model that he sought to advance in response to the apparent sterility of logical positivism. David Hull's *Science as Process* (1988), on the other hand, advances the contrary argument that evolutionary modes of explanation provide a powerful model for understanding the growth of knowledge. Finally, Helen Longino's *The Fate of Knowledge* (2002), in a manner similar to Hull but animated by very different motives, proposes that the epistemological and methodological pluralism of contemporary biological thought offers scope for the development of a less hierarchical understanding of the relationship between knowledge and power.

Seeking to disentangle these two such important approaches to the relationship between biology and modern philosophy, Eugene Thacker (2005) draws a very useful set of distinctions between the philosophy of biology and biophilosophy. Thacker regards the philosophy of biology as engaged in the identification of the principle of life and the boundaries of its articulation, as well as evolving along two related axes. The first of these two axes involves the conceptual analysis of characteristics that are understood to define the organism's essence or organizing principle. The second axis concerns instead the historical development of such understanding. True to its roots within the analytical tradition, Thacker adds, the philosophy of biology can be understood as posing the ontological question about the difference between that which is living and that which is not living by reframing the question as an epistemological one about what makes biology different from other fields of study, such as chemistry and physics. Claude Bernard's *Introduction to the Study of Experimental Medicine* (1865) testifies to the long history of such conjunction of epistemology and concern to characterize life as a material phenomenon.[60] Biophilosophy, on the other hand, seeks to resist the reduction of the ontological question about life to epistemological arguments, so that instead of articulating a conception that would adequately describe the essence of life, it aims to articulate those processes involved in the continuous transformation and diversification of life. Thus, whereas the philosophy of biology is increasingly concerned about the reduction of life to determinate quantities and related questions about essential material constituents, biophilosophy seeks to characterize the mechanisms involved in recombination and proliferation beyond all determinations, including biological determinations. In other words, as Thacker puts it, 'whereas the philosophy of biology proceeds by the derivation of universal characteristics for all life, biophilosophy proceeds by drawing out the network of relations that always take the living outside itself'. As such, if biophilosophy can be said to seek the renewal of vitalism in order to purge biology of all essence, the philosophy of biology seeks constantly to purge biology of all vitalism, the understanding that the ever proliferating heterogeneity of the phenomenal world attests to some immanent power that is, however, irreducible to substance and extension.

There is much about Thacker's account of the relationship between biophilosophy and the philosophy of biology that is very useful to any examination of the ontological assumptions and epistemological practices involved in understanding the relationship between life, death and truth today. At the same time, as James Griesemer and John Protevi's shared interest in the abstractions involved in the notion of genetic determination would seem to attest, the account overlooks how August Weismann may be the point of encounter between these two so very different modes of thinking about the relationship, testing the limits of both modes of juxtaposing biology and philosophy, by asking repeatedly what is the status of the organism, both centre of reflection and product of historical processes in which it is but an insignificant part. There seems to be something irreducibly important about the organism, however impossible it may be to separate the organism from genealogical and ecological contexts. If Deleuze extends an invitation to see entities such as the organism as inseparable from the problems they are called upon to resolve, they are not to be conflated with these same problems. However obscurely, if not darkly, these pivotal figures would seem to actualize the boundary between the molar and molecular scales, and this boundary would seem to coincide with death.[61]

Finally, while thinking about the boundary between life and death is a matter for the next chapter, the references to mortality and darkness resonate with Foucault's observation, in *The Order of Things*, that:

> [M]an has not been able to describe himself as a configuration in an episteme without thought at the same time discovering, both in itself and outside itself, at its borders yet also in its very warp and woof, an element of darkness, an apparently inert density in which it is embedded.[62]

Such obscurity would seem to be the very condition of possibility of thought and its endless renewal.[63]

CHAPTER 7

The arts of living and dying

The previous chapter sought to draw out the differences between Michel Foucault's and Gilles Deleuze's understanding of the relationship between life, death and philosophy. Much attention was devoted to Deleuze's view that Foucault had attached excessive importance to human mortality, seeking to clarify Deleuze's argument by situating it between Henri Bergson's and August Weismann's understanding of life, death and embodied existence. In so doing, the chapter drew out how the difficulties involved in defining the relationship between the human subject and Deleuze's pivotal notion of a body without organs are not dissimilar to those confronting the attempt to offer a positive representation of the organism and its relationship to mortality within the domain of biology and medicine. As the like of Michel Serres (1980) might suggest, however, the urge to differentiate and clarify can deflect attention away from both the proximity of the most antithetical positions and the fundamental importance of their entangled relationship.

In an important essay on Foucault's and Deleuze's understanding of human existence, Giorgio Agamben (1999) observes how Deleuze offers a wholly different way of understanding such existence, beyond power and knowledge.[1] Seeking to explain, Agamben returns to Deleuze's formative reading of Baruch Spinoza and to the figures that Deleuze employs to convey his distinctive understanding of life as the process of perpetual differentiation. He discusses how Deleuze refers his readers to the scene in Charles Dickens's *Our Mutual Friend* (1865) where Riderhood, one of the novel's main characters, is brought out of the Thames. Deleuze writes:

> No one told better than Dickens what a life is, taking account of the indefinite article as an index of the transcendental. At the last

minute, a scoundrel, a bad subject despised by all, is saved as he is dying, and at once all the people taking care of him show a kind of attention, respect, and love for the dying man's smallest signs of life. Everyone tries to save him, to the point that in the deepest moment of his coma, the villainous man feels that something sweet is reaching him. But the more he comes back to life, the more his saviours become cold, and he rediscovers his coarseness, his meanness. Between his life and his death there is a moment that is nothing other than that of a life playing with death. The life of the individual gives way to an impersonal yet singular life, that is, of the subjectivity and objectivity of what happens. 'Homo tantum', for whom everyone feels and who attains a kind of beatitude.[2]

Deleuze, according to Agamben, is captivated by the conjunction of singularity and the universal, what Deleuze names elsewhere 'the movements of the infinite'.[3] Agamben goes back to Dickens's text to argue that there is much more at stake in this scene. Dickens's text reads:

> See! A token of life! An indubitable token of life! The spark may smoulder and go out, or it may glow and expand, but see! The four rough fellows seeing, shed tears. Neither Riderhood in his world, nor Riderhood in the other, could draw tears from them; but a striking human soul between the two can do it easily. He is struggling to come back. Now he is almost there, now he is far away again. Now he is struggling harder to get back. And yet – he is instinctively unwilling to be restored to the consciousness of this existence, and would be left dormant, if he could.[4]

Agamben claims that the flickering spark animating Riderhood, he who is precariously suspended between life and death, is no less than a figure for 'bare biological life as such'.[5] As Melinda Cooper observes, Agamben's further claim that this figure also instantiates Spinoza's understanding of immanent cause, in all its splendour and 'complete beatitude', is questionable because, as Cooper puts it, 'for Spinoza, life is never a predicate that can be attributed to (or separated from) a finite potential substance but rather infinite substance itself, of which there is only one'.[6] There can be little doubt that Deleuze's reading of the scene is truer to Spinoza's insistence on the continuity of substance, and that Agamben's attests to the manner in which Agamben, like Foucault and Heidegger before him, will not dissociate life and the organism. Significantly, Cooper also seems to share Mitchell Dean's preoccupation about the ethics of contemporary reflections on the organization of biopolitical order.[7] She suggests that these differences of understanding between Agamben and Deleuze are of some importance to any critical analysis of the profound division of attitude toward contemporary biomedical developments between those who fear the reduction of the subject to the most denuded and basic form of life, and those who regard the detachment of life and body in a manner that is never very far removed from the promissory economy of advanced consumer capital. The contrast is very useful, but it also is unclear whether these two positions are quite as incompatible as such contrast would seem to imply.

Despite all the important differences outlined in the preceding chapters, this last substantive chapter seeks to move in the opposite direction, by exploring further the points of commonality between Agamben, Foucault and Deleuze. It will do so by examining first the complexity of those situations where contemporary biomedical developments succeed in blurring the boundaries between life and death just as in the scene that Dickens sets in *Our Mutual Friend*. The remainder of the chapter will then seek to bring the discussion to a close by examining how the confrontation with mortality, not as biomedical or philosophical abstraction, but as impending prospect, helps to draw out the proximity of Deleuze's, Foucault's and Agamben's understanding of the relationship between life, death and freedom.

Between life and death

In *The History of Sexuality* (1976), Michel Foucault discussed how the modern deployment of governmental powers focused increasingly on the human subject in his or her existence as an embodied entity. As was observed in the last chapter, Foucault remarked how:

> For millennia, man remained what he was for Aristotle: a living animal with the additional capacity for a political existence;

modern man is an animal whose politics places his existence as a living being in question.[8]

The ominous connotations of this biological redefinition of political subjectivity, which are signalled by the words 'modern man is an animal whose politics places his existence as a living being in question', are articulated more fully, and problematically so, in the course lectures that Foucault was delivering simultaneously at the Collège de France. In the last of these lectures, Foucault described how the maximization of the modern political collective's productivity came to require 'purification', or, in other words, the elimination of all that which might contaminate the biopolitical collective. The classical, sovereign power to punish infraction of the law by killing the transgressor thus resurfaces as a supplementary power of the modern, biopolitical governmental apparatus, this time taking the form of 'race war'. Foucault also argues that this alignment reached its 'paroxysmal' point under German national socialism, on the eve of its final defeat, when the remaining citizens of the defeated German nation were expected to die because they had failed the ultimate test of their racial superiority.[9]

In a widely acclaimed response to Foucault's understanding of the relationship between governmental powers and embodied existence, Giorgio Agamben (1995) rejects the Foucauldian construction of the relationship between biopolitical techniques of governance and sovereign power. He does so by wending his way from the *arcana* of Classical jurisprudence to novel conjunctions of jurisprudence and medical reason that were forged during the early twentieth century, and from there to those increasingly mundane scenes that Foucault had sought to capture in his discussion of Francisco Franco's terminal coma. During the course of the same lecture just discussed, Foucault had also observed how 'the death of Franco [is] a very, very interesting event':

> It is very interesting because of the symbolic values it brings into play, because the man who died had, as you know, exercised the sovereign right of life and death with great savagery, was the bloodiest of all the dictators, wielded an absolute right of life and death for forty years, and at the moment when he himself was dying, he entered this sort of new field of power

over life which consists not only in managing life, but in keeping individuals alive after they are dead. And thanks to a power that is not simply scientific prowess, but the actual exercise of the bio-political power established in the eighteenth century, we have become so good at keeping people alive that we've succeeded in keeping them alive when, in biological terms, they should have been dead long ago. And so a man who had exercised the absolute power of life and death over hundreds of thousands of people fell under the influence of a power that managed life so well, that took so little heed of death, and he didn't even realize that he was dead and was kept alive after his death.[10]

In a manner consistent with the account of Foucault's understanding of the relationship between life and death that was advanced in the previous chapter, Foucault here privileges life over death repeatedly, but in a manner such that the latter seems so absolutely refractory to human intervention that it could stand for that which is true and unquestionable, always gnawing away at embodied existence. Thus, although Foucault was prepared to concede the intelligibility of the notion that, 'from the point of view of life and death', the position of the subject might be regarded as 'neutral', he regarded such symmetry as no more than a 'theoretical' possibility.[11] This understanding is diametrically opposed to that which Agamben seeks to offer in his own reflections of those situations consequent upon modern medicine having, as Foucault put it, 'become so good at keeping people alive that [it has] succeeded in keeping them alive when, in biological terms, they should have been dead long ago'. After discussing the development of brain death as an alternative to cardio-vascular definitions of death, Agamben writes:

> It is not our intention to enter into the scientific debate on whether brain death constitutes a necessary and sufficient criterion for the declaration of death or whether the final word must be left to traditional criteria. It is impossible, however, to avoid the impression that the entire discussion is wrapped up in inextricable logical contradictions, and the concept of 'death', far from having become more exact, now oscillates from one pole to the other with the greatest indeterminacy, describing a

> vicious circle that is truly exemplary. On the one hand, brain death is taken to be the only rigorous criterion of death and is, accordingly, substituted for systematic or somatic death, which is now considered insufficient. But on the other hand, systemic or somatic death is still, with more or less self-consciousness, called in to furnish the decisive criterion.[12]

Agamben then goes on to discuss the juridical consequences of such ambiguity and uncertainty, writing that:

> A perfect example of this wavering is the case of Karen Ann Quinlan, the American girl who went into deep coma and was kept alive for years by means of artificial respiration and nutrition. On the request of her parents, a court finally allowed her artificial respiration to be interrupted on the grounds that the girl was to be considered as already dead. At that point Karen, while remaining in a coma, began to breathe naturally and 'survived' in a state of artificial nutrition until 1985, the year of her natural 'death'. It is clear that Karen Quinlan's body had, in fact, entered a zone of indetermination in which the words 'life' and 'death' had lost their meaning.[13]

Basically, situations such as this lead Agamben to conclude, *contra* Foucault, that, ever since the human was first understood as an animal endowed with the additional capacity to become a political subject, the very understanding that Foucault regards as effectively immemorial, the articulation of political sovereignty has rested on the prior assumption of a form of being that is neither dead nor alive, except by virtue of juridical determination. Agamben labels this form of being that is neither dead nor alive as bare life. More importantly still, toward the conclusion of his argument, Agamben returns to contemporary biomedical confrontations with the boundaries between life and death, to distil their fullest implications. He writes:

> We enter the hospital room where the body of Karen Quinlan or the over-comatose person is lying, or where the neo-mort is waiting for his organs to be transplanted. Here biological life – which the machines are keeping functional by artificial respiration, pumping blood into the arteries, and regulating the

blood temperature – has been entirely separated from the form of life that bore the name Karen Quinlan: here life becomes (or at least seems to become) pure *zoē*. When physiology made its appearance in the history of medical science toward the middle of the seventeenth century, it was defined in relation to anatomy, which had dominated the birth and the development of modern medicine. And if anatomy (which was grounded in the dissection of the dead body) was the description of inert organs, physiology is 'an anatomy in motion', the explanation of the function of organs in the living body. Karen Quinlan's body is really only anatomy in motion, a set of functions whose purpose is no longer the life of an organism. Her life is maintained only by means of life-support technology and by virtue of a legal decision. It is no longer life, but rather death in motion. And yet since life and death are now merely bio-political concepts, as we have seen, Karen Quinlan's body – which wavers between life and death according to the progress of medicine and the changes in legal decisions – is a legal being as much as it is a biological being. A law that seeks to decide on life is embodied in a life that coincides with death.[14]

On this understanding of the biopolitical subject as suspended between life and death, subject to medico-juridical decision, that which, to Foucault, was the suicidal, paroxysmal point of modernity becomes the paradigm of the modern governmental formation. If, under national socialism, every embodied subject found himself or herself precariously suspended between life and death, this was no historical exception, but the intimation of the coming community in which the state of exception is disclosed as the rule of governance. As Agamben puts it, as he brings the argument to a close, and responding directly to Foucault's own, more affirmative closing of *History of Sexuality*:

> The 'body' is always already a bio-political body and bare life, and nothing in it or the economy of its pleasure seems to allow us to find solid ground on which to oppose the demands of sovereign power.[15]

Without a doubt, there is much substance to Agamben's diagnosis of the contemporary convergence of medical and juridical formations, but the situation also is considerably more complicated than Agamben's account might convey.

In *Twice Dead* (2002), Margaret Lock notes how the concept of brain death, which so exercises Agamben, effectively reverses the pre-modern relationship between biological and social death. Today, the anthropological person is deemed to be legally, if not socially, dead when their brain ceases to function, but biomedical technology can keep the rest of the embodied individual biologically alive almost indefinitely, or at least until transplant surgeons can remove their still-living organs. If this is currently a phenomenon of limited historical importance, the same cannot be said of the 'living death' experienced by increasing numbers of incapacitated people whose survival is tenuously guaranteed by their attachment to life support technology and other forms of intensive care, specifically the terminally ill and the elderly, as the latter approach the end of their increasingly medicalized lives. This situation is resulting in the proliferation of discourses about the easy death, the good death and, most basically, the right to die. When the terminally ill and elderly are denied this last right because it supposedly is always better to be alive than dead, and they then reply that they are already 'dead', this is no longer a matter of histrionics, but a cause for advancing a 'politics of death'. On the medical side, this situation is translated into the growth of palliative medicine as a specialization that seeks to rejoin social and biological death by desisting from any attempt to prolong life needlessly, that is, by desisting from any attempt to prolong biological life beyond any hope of return to social life.[16] If this approach to the management of the boundary between the life and death of the individual is in practice quite problematic, the medical profession, at least in the United Kingdom, adamantly refuses to sanction the adoption of any measures that would intentionally accelerate the advent of death. It would seem that the great majority of the lay public has no such qualms. It not only denounces the hypocrisy of the medical profession, as the latter plays with intangible notions of intentionality to legitimate the administration of drugs to relieve pain that in fact accelerate the advent of death, but it also increasingly demands the legalization of assisted suicide. The accompanying arguments about medical practitioners' actual intentions and legislators' insistence that, if allowed, such assisted suicide should be left in the hands of the medical profession, betray an inordinate confidence in the profession's understanding of the threshold between life and death, which is in fact unwarranted. The net effect of all these ambiguities and complications is that the concept of the 'sanctity of life', the imperative to protect life as much as is humanly possible, which would appear to have long organized religious as well as medical discourse, is steadily being supplanted by the concept of 'quality of life', whereby it must now be recognized that death is sometimes preferable to a life reduced to bare biological function, to bare life. Bare life, in other words, is not the fruit of anyone's overwrought and dystopian imagination, but produced and reproduced in the most mundane contemporary confrontations with human mortality.[17]

Against this background, a number of scholars have drawn on Gilles Deleuze and Deleuzian thought to emphasize how the demands that assisted suicide should be legalized must be viewed as the refusal to submit to biopolitical order and the final fulfilment of the truly autonomous and creative individual, in the very act of its final dissolution. Thus, Patrick Hanafin, focusing particularly on the situation in the United States, writes:

> In the case of assisted suicide, we can see the play between the micro-politics of individuals who are attempting to self-style their deaths and public officials, in the form of judges, who attempt to maintain the status quo and prevent the creation of this new right. It is the beginning of an elaboration of a new right, an opening to a new way of becoming indiscernible.[18]

Hanafin concludes that opting to wrest the timing of one's death from the control of medical and juridical institutions, like the agreement to donate one's organs and other bodily parts, must be regarded as a form of 'self-artistry'.[19]

The situation, in sum, is considerably more complex than Agamben will allow. This said, while it is unclear what is Agamben's view on suicide as a form of opposition to 'the demands of sovereign power', the next two sections will seek to examine more closely the plausibility and consistency of this understanding of suicide as a form of 'self-artistry' with Deleuze's own understanding of the relationship between life, death and human existence.[20]

The art of life

Paul Rabinow is an acute observer of contemporary developments in the biomedical sciences and their implications for any understanding of what it means to be human today. Rabinow's reflections on the convergence of art and the biomedical sciences can serve usefully to draw out the ambiguities of the embrace of death as a form of affirmative, artistic endeavour.

As observed previously, Michel Foucault, who in many ways can be regarded as Rabinow's teacher, proposed that modernity was characterized by two distinct conjunctions of knowledge and power, the anatomo-political and biopolitical formations.[21] Rabinow would also observe, however, that the convergence of biomolecular and biomathematical considerations involved in propositions such as those advanced by contemporary biogerontology blur the distinction between the two formations, thus calling into question the contemporary significance of Foucault's otherwise persuasive proposition. Exercised by this question, Rabinow turns to Gilles Deleuze's argument that Foucault had failed to delineate what form of life would follow upon the death of Man because he did not recognize how life was breaking free from the equation of the organism and its life course from birth to death. Rabinow concurs, but, as observed earlier, he is averse to any metaphysical speculation and he thus suggests in more historically circumscribed fashion that contemporary social organization should be understood as structured increasingly by interventions at a biological level and in a manner such that the modern attempt to reduce social relations to a matter of biology gives way to recombination as the operative principle. As such, the contemporary biomedical laboratory should be understood not as the centre of biopolitical calculation, but as the site of experiments in the creation of new problems and challenges. This, as Rabinow puts it in his reflections on French responses to contemporary biomedical developments, requires an alternative comportment to the modern hermeneutic of suspicion, the search for some hidden logic and explanation, which the like of Jacob Taubes, Giorgio Agamben's touchstone, refers back to the Gnosticism of Judeo-Christian culture. Rabinow's answer is to learn from the biomedical scientists themselves and to adopt instead 'an experimental mode of inquiry … where one confronts a problem whose answer is not known in advance rather than already having answers and then seeking a problem'.[22] Significantly, this notion seems to lie at the very heart of Rabinow's long-standing disagreement with the very same Agamben and to have animated his comment that Agamben's arguments about the nature of contemporary biopolitical governmentality entail 'highly general philosophical deployments of ... terms which are totalising and misleading'.[23]

Seeking to articulate the experimental comportment further, Rabinow describes in *Anthropos Today* (2003) the response of visual artists of the late nineteenth and early twentieth centuries to the introduction of industrially manufactured, standardized palettes, focusing especially on Marcel Duchamp.[24] Duchamp's iconoclasm, Rabinow argues, rested not so much on his use of these industrially manufactured palettes, or his move away from painting, as it did on his understanding that the very nature of art was transformed by this technological innovation and that the challenge was not to flee the present, but to work differently within the present. Importantly, Rabinow suggests that, in so doing, Duchamp refused not just the conservative flight into the past of tradition, but also the futurism of the avant-garde, because both seek to evoke some other, redeemed time. Something very similar is at issue today in bio-art. Bio-art involves a redefinition of art as mobilizing biomolecular components and associated practices to construct something like a phenomenology of life itself. The issue is not how best to draw attention to the artefacts of contemporary biotechnology and to the future they supposedly hold in store, but to convey an understanding of these products as part of a complex and historically very specific process of production. Importantly, in so doing, bio-art partakes in a demystification of biotechnology. If the artisan's workplace once was the laboratory, but the word is now identified with a particular kind of labour that aims at disclosure, this is not simply due to appropriation by scientific institutions, but also by virtue of the simultaneous transformation of the artisan into artist who contemplates truth. Bio-art, however, reconstitutes the laboratory as the site of creation, thus not only calling into question received notions of the artist as the contemplative subject of truth, but also collapsing the contemporary differentiating and dividing vocabulary of artistic contemplation and scientific disclosure because both are involved in creating new and unprecedented forms of life.

Significantly, while Taubes's diagnosis of the present historical moment, as discussed in earlier chapters, is very different to Rabinow's, bio-art can also be regarded as the site where their views might be said to converge.[25] On the one hand, Rabinow maintains that bio-art must be understood as partaking in a restatement of Hans Blumenberg's arguments about the legitimacy of the modern age, the notion that the structure of modern culture is novel and should not be understood as shaped by past questions about the order and meaning of the phenomenal world. Bio-art transfers the debate into a new, biotechnological and biosocial context, and, in so doing, transforms the argument. On the other hand, Rabinow's discussion of Gerhard Richter's figurative reflections on the task of representation seem to suggest that bio-art, if not the artefacts of contemporary biotechnology themselves, should be regarded as a forms of surrealist endeavour because what is at a stake is the possibility of creating life forms with no natural counterpart, that is, the possibility of creation beyond all imitation.[26] It is then noteworthy that Taubes was much taken by Louis Aragon and other surrealists' understanding of poetry and its task, the creation of something 'beyond' that which exists, either materially or ideally.[27] Taubes was struck in particular by the disjunction between these surrealists' uncompromising materialism and atheism, on the one hand, and, on the other hand, their interest in playing with the relationship between sign and referent, their immersion in the excessiveness of linguistic representation. He thus came to regard the surrealist understanding of poetic creation as the heir to Gnosticism. As he put it, passing through the historic transfer of the site of reflection from the community to the soul and spirit,

> Gnostic a-cosmism and the worldlessness of modern poetry is based on the fact that in the protest and provocation of Gnosticism as much as in the protest and provocation of modern poetry the pre-given interpretation of the world as a whole is bracketed. The Gnostic doctrine of redemption is a protest against a world ruled by *fatum* or by *nomos*. This *fatum* present itself in the mythological style of Gnosticism as personified powers: astrological determinism. The world as it is represented by the interpretation of modern science and technology against which modern poetry turned in varying phases since

romanticism, regains a mythical coherence as a unified whole: natural-scientific determinism. The poetic protest turns against the enslavement to nature of science and technology, the consequence of knowledge as power that can be wielded only in the form of domination and coercion of a demystified nature.[28]

Surrealism, in other words, is to be understood as the heir to Romanticism, and the appeal to something beyond phenomenal appearance as pointing to the perception of something shared with, but beyond, the claims of science and technology. True to such heritage, surrealist poetry, as Taubes puts it, 'no longer copies an exemplary creation, the order of the world, [but] rather it disassembles and destroys this order, in order to create out of the depths of the soul a new world from its individual parts'.[29] When Richter states that he does not 'wish to imitate a photograph', but wishes instead 'to make one', he would seem exemplify this endeavour to take that which is and make of it something wholly different and without precedent.[30] This understanding of poetic creation is precisely Henri Bergson's understanding of evolutionary process, which Deleuze then seeks to reaffirm in *Difference and Repetition* (1968).

If bio-art would therefore seem to articulate an understanding of the present historical conjunction whereby radically different perspectives on human existence converge, it remains unclear how this understanding is to be translated into norms of conduct adequate to the contemporary biopolitical context. Importantly, while seeking to corral many of the philosophers considered thus far and articulate the contours of an affirmative form of existence, Tim Campbell (2011) evokes a comportment that would seem to be the heir to the complex genealogy of bio-art just outlined. Campbell takes his cue from Foucault's contrast between Classical and Judeo-Christian modes of constituting the self. In another set of lectures at the Collège de France, contemporaneous with the production of *Care of the Self* (1984), Foucault suggested that:

> If we want to understand the form of objectivity peculiar to Western thought since the Greeks we should maybe take into consideration that at a certain moment, in certain circumstances typical of classical Greek thought, the world became the correlate of a *tekhnē* – I mean that at a certain moment it ceased

being thought and became known, measured, and mastered thanks to a number of instruments and objectives which characterized the *tekhnē*, or different techniques – well, if the form of objectivity peculiar to Western thought was therefore constituted when, at the dusk of thought, the world was considered and manipulated by *tekhnē*, then I think we can say this: that the form of subjectivity peculiar to Western thought ... was constituted by a movement that was the reverse of this. It was constituted when the *bios* ceased what it had been for so long in Greek thought, namely the correlate of a *tekhnē*; when the *bios* (life) ceased being the correlate of a *tekhnē* to become instead the form of a test of the self.[31]

In other words, if the contemporary, biopolitical ordering of the polity seems inevitably to entail some discrimination between forms of life worth living and those not worth living, this is because this ordering is predicated on an understanding of the subject inherited from Judeo-Christian culture. As a result of this inheritance, the body has ceased to be the site of self-construction and has come to be regarded instead as the site of discrimination between proper and improper modes of conduct, and the development of such capacity for discrimination is required of all those who would accede the good life. As such, the constitution of the good life must also rest upon the mobilization of some negative determination as the purifying agent, death being the paramount such agent. While trying to specify how this insight might be mobilized to overturn the relationship, Campbell seeks to evoke the possibility of a more playful relationship between self and the body. Drawing on both Deleuzian notions about the plasticity of life and Walter Benjamin's notions about children's extraordinary ability to overcome the 'dour naturalism' of toys and become the very things their toys would imitate, Campbell writes that 'a *tekhnē* of *bios* thought through play might be one yet unexplored way to forgo "the dour naturalism" of biopolitics today, in which the object of politics would be merely biological life or that would have the object of life be thinkable only as part of a negative politics'.[32] Bearing in mind the importance of these notions of playfulness to Agamben, Campbell would thus seems to reaffirm the occasional proximity between Agamben and Deleuze.[33] More importantly, however, this notion also recalls Stelarc, the performance artist who has invested much in the notion of the body as a field of playful de-composition and re-composition, and his, her or its own evocation of a future human who will be endowed with photosynthetic skin:

> With such a skin we would no longer have need of a mouth to chew, of a throat to swallow, of a stomach to digest, of lungs to breathe. We would be able to leave the human and replace useless organs with technologies. Ha-ha-ha-ha-ha-ha-ha-ha-ha-ha-ha-ha![34]

Not only is it unclear to what the 'we' evoked here refers, and not only will anyone sceptical about the transformative power of the avant-garde hope that tongues and cheeks are lodged very firmly together, but Stelarc's performance also evokes all the ambiguities of any playful relationship between self and the body.

Stelarc lends material substance to a post-humanist critique of the human form, but by so rehearsing the preoccupation with the limits of the self, she, he or it also accentuates the closure of the self on itself, which, as Deleuze puts it, in his discussion of sexuality and its development, amounts to a denial of the body without organs by closing upon 'a full depth without limits and without exteriority'.[35] In other words, Stelarc's promised creature is a monstrous one. Perhaps we have not then moved any great distance from either the Romantic heyday of biology and its own proliferation of monstrous figures on the boundaries of poetry and biology, or from H. G. Wells's reconfiguration of that heritage in the like of *The Island of Doctor Moreau* (1896). At the same time, however, like all monstrous figures, Stelarc highlights boundaries and all the ambiguities that would seem to proliferate on these sites of differentiation.[36]

Death, dying and the reproduction of biopolitical order

Returning to the notion that wresting the timing of one's death from the control of medical and juridical institutions might be regarded as a form of self-artistry, the ambiguities of bio-art with respect to the relationship between self and the body help to better understand the complexities involved in recent debates over the end of life and the call for the relaxation of the laws and regulations governing this moment in the course of human existence.

The contemporary prominence of discourses about an easy death, a good death and the right to die leads Clive Seale and others to contest Philippe Ariès's formative thesis about the historical disappearance of death. Seale calls this understanding into question by observing that,

> In a manner similar to the mortuary rituals described by anthropologists … [contemporary mortuary rituals] … offer both dying people and the bereaved an opportunity for organised resurrective practice, whereby participants engage in relatively explicit affirmations of membership in the human social bond, in the face of its destruction by the death of a member.[37]

The terms of the debate about Diane Pretty's request that her husband be allowed to help her commit suicide, which gripped the British public's attention between 2001 and 2002, suggest however that the resemblance of contemporary and pre-modern discourse that Seale thus advances should be regarded as formal, rather than substantive.[38] Pre-modern ritual of dying was oriented toward, and enjoined continuous reflection upon an end, namely salvation on the Day of Judgement. As such, the ritual aimed to return the original, sacrificial 'gift of life' and thus realize the historical promise enacted by this founding sacrifice. Such symbolic exchanges, however, are no longer meaningful. While the pervasiveness of discourse about dying and accompanying rituals is undeniable, what is more striking is how discussion on the meaning of death itself is noticeably absent. Consequently, contemporary ritual seems more a discursive performance about an empty centre, so that noun and verb merge and death becomes dying. Such historical transformation of the noun, the reduction of content to form and performance of form, brings some intelligibility to the paradoxical concomitance of denunciations of all the residual attachments to Judeo-Christian values, such as notions about the sanctity of life, that stand in the way of assisted suicide, and the public celebrations of Pretty's death with the words 'she is free at last'.[39] This celebration begs so many questions about what is freed, freed from what and in the name of what truth, that this situation is perhaps best regarded not so much as a contradiction, occasioned by the incomplete secularization of contemporary society, but as a symptom of evacuation.

Against this background, it is worthwhile observing how Pretty's case could be summarized as a clash between two fundamentally different conceptions of the relationship between life and death, which, as observed earlier, are usually characterized as commitment to either the sanctity of life or the improvement of the quality of life. At the same time, many aspects of the case cast doubt on such a seductive construction.

The Law Lords, who were called upon to decide whether to sanction Pretty's request or not, were moved by Pretty's difficult situation, facing as she did the prospect of accelerating degeneration and eventual death. They maintained however that they were bound to uphold the sanctity of life. If Pretty's condition made it impossible for her to bring her life to an end, neither she nor any legal institutions enjoyed the right to control the manner and timing of her death. Death, in other words, was a sovereign power. Yet, when previously asked to consider the case of Tony Bland, the Law Lords had agreed to allow the withdrawal of the medical support that had kept Bland alive for four years, in a persistent vegetative state. Bland died promptly thereafter. According to the Law Lords, this course of action did not constitute murder because, even if the withdrawal of medical support was bound to result in Bland's death, there was no intention to kill Bland. If there is any point of agreement between some of the most thoughtful proponents and opponents of euthanasia and assisted suicide, Peter Singer and John Keown, it is that the decision regarding Bland was not based on any principled defence of the sanctity of life.[40] With regard to Pretty, the Law Lords were in fact quite capable of determining where the boundaries between life and death should be drawn and they did so on the basis of pragmatic considerations. They referred Pretty and her lawyers to the Select Committee on Medical Ethics and its own prior expression of concern about any relaxation of existing legislation governing the management of end of life:

> We are also concerned that vulnerable people – the elderly, lonely, sick or distressed – would feel pressure, whether real or imagined, to request early death. We accept that, for the most part, requests resulting from such pressure or from remediable depressive illness would be identified as such by doctors and

managed appropriately. Nevertheless we believe that the message which society sends to vulnerable and disadvantaged people should not, however obliquely, encourage them to seek death, but should assure them of our care and support in life.[41]

Significantly, the Law Lords themselves reinforced the point by writing that: 'it is not hard to imagine that an elderly person, in the absence of any pressure, might opt for a premature end to life if that were available, not from a desire to die or a willingness to stop living, but from a desire to stop being a burden to others'.[42] It is not clear what prompted this further observation, but contemporaneous debates over revelations about bedside instructions not to resuscitate hospitalized patients highlighted how the elderly were threatened most by changing attitudes toward the determination and management of the end of life.[43] Whatever the specific historical reasons may have been, the Law Lords thus appeared to regard their responsibility toward the life of those who might be less capable of resisting unwarranted, external pressure to bring their lives to an end as outweighing Pretty's preferences. The Law Lords' position was, in sum, utilitarian and consequentialist, rather than principled.[44]

Pretty's position was no more consistent and principled. According to Pretty and her lawyers, existing legislation failed to recognize how contemporary biomedical technology was bringing about a situation where the paradoxical notion of a living death was becoming a mundane reality. Being dead or alive was not a matter of categorical opposition, but part of a continuum whereby death might sometimes be preferable to life. Pretty had become so completely dependent on others, human and non-human, that, when asked whether life was not better than death, she would sometimes answer that she was already dead. Basically, she wanted to regain the control over the time and manner of her death lost to others. As she put it in one of her many interviews: 'If I am allowed to decide when and how I die, I will feel that I have wrested some autonomy back'.[45] Consequently, Pretty and her lawyers insisted that the refusal to acknowledge her freedom to choose between life and death, and to sanction the mechanisms that would realize such freedom amounted to infringement of her 'human dignity'. Strikingly, to advance this powerful argument, the vulnerable woman who had no choice but to live 'in the thrall of machines' took advantage of another machine to publicly downplay the relief offered by palliative care and emphasize instead how the law was forcing her to endure insufferable agony, and how, under such conditions, her agony could only end with her drowning in her own saliva. This and other similar images allowed Pretty and her lawyers to mobilize the public and the media to understand the case as an unambiguous confrontation between an outmoded principle, the sanctity of life, and its horrific consequences. For Sarah Barclay, who was involved in the production of a BBC documentary on the case, the law was forcing Pretty to live like a 'tortured animal'. As Barclay put it:

> When Diane Pretty wants to tell you something, she makes a noise that is a cross between a grunt and a moan … If no one can understand what Diane is trying to say, the grunt becomes a low-pitched scream. If you still cannot understand, it starts to get louder. Eventually, she opens her mouth wide and howls, a sound which makes the hairs stand up on the back of your neck. You can hear this cry from half-way down her street. If you did not know that Diane Pretty lived in that house, you would think it was an animal being tortured.[46]

A personal preference for the timing and manner of one's death was thus transformed into a public campaign for the legalization of assisted suicide. Ironically, this was such a powerful story about the evils of sovereign power that, when some journalists insisted on describing Pretty as someone completely dependent on others' whim, Pretty would sometimes have to remind them that she was not nearly so powerless. As such, Pretty's success in pursuing her case as far as she did depended on very complex, even artful, choreographies of strength and vulnerability, and a good deal of confusion of boundaries between humans, animals and machines.[47]

In sum, every position articulated during the eleven months between Pretty's first request that her husband be allowed to aid her suicide and her death was intensely historical and untrue to any categorical distinction, but the situation was undoubtedly productive, reinforcing the culturally dominant notion of the wholly autonomous subject, free from the grip of all power.

At the same time, John Protevi's analysis of the controversies surrounding the termination of the medical care keeping Terri

Schiavo alive, but in a persistent vegetative state, suggest that it is not just the endeavours to wrest control of the timing of death from the hands of the medical profession, but also the very analysis of such situations, that can fall captive to the reproduction of biopolitical formations.

Protevi (2008) suggests that Gilles Deleuze provides a more productive approach to the analysis of those moments when human existence is suspended precariously between life and death than either Giorgio Agamben or Michel Foucault because neither of the last two can provide the positive and affirmative account of embodied existence necessary to understand the emotional intensity of such situations. Such situations, according to Protevi, call for the identification of an ethically charged form of existence on the boundary between life and death that lies beyond either the sovereign power of the state or the subjectivity constructed by the more dispersed disciplinary powers of medicine, law and the media. What is required is a fully relational understanding of all parties involved that will allow for the call to sympathy that issues from such cases and yet also allow the elaboration of normative criteria that will ensure the empowerment of the objects of such sympathy. Protevi seeks to articulate the contours of such an alternative by focusing on the relationship between Schiavo, Schiavo's husband and Schiavo's parents. The struggles between husband and parents over the suspension of the medical care that kept Schiavo suspended between life and death, he suggests, actualize a nexus wherein the required singular existence might be glimpsed. Protevi then cites Deleuze and Félix Guattari's *A Thousand Plateaus* (1980) to convey some understanding of the uniqueness of such nexus:

> Every love is an exercise in depersonalisation on a body without organs yet to be formed, and it is at the highest point of this depersonalisation that someone can be named, receives his or her family name or first name, acquires the most intense discernibility in the instantaneous apprehension of the multiplicities belonging to him or her, and to which he or she belongs.[48]

For all the affective power of such representation of love and its shattering intensity, it is unclear how this can possibly answer the normative requirements of jurisprudence and all that flows from such requirements. More importantly, it would also seem that

Protevi overlooks how the relational nexus he constitutes comes at the cost of other relationships, as it must always do, because, as Deleuze and Guattari also observe, 'the distinction to be made is … between different types of multiplicities that coexist, interpenetrate, and change places – machines, cogs, motors, and elements that are set in motion at a given moment, forming an assemblage of productive of statements: "I love you" (or whatever)'.[49] Protevi would seem to thus privilege familial attachments over and above the relationships forged by the powers of medicine, law and the media. It is not clear why the former set of relationships, which share much with those that are secured by processes of filiation, are so preferable to other powers, and, if anything, the privilege accorded to them would seem to reinforce the contemporary mode of biopolitical organization.[50]

In sum, if death is the signature of sovereign power, Pretty's and Schiavo's cases suggest that the investment in securing control of the time and manner of a loved one's or one's own death would seem inescapably to redouble the hold of biopolitical formations, and that contemporary neo-Deleuzian critical practices would seem to be bound up in such redoubling.[51] There is no escape from the grip of death and power.

Life, death and freedom

Much of *Biopolitics and the Philosophy of Death* has been taken up with the attempt to make sense of the present biopolitical moment and all its complexity by differentiating and clarifying the manifold implications of the terms thus differentiated. The remainder of this chapter seeks to bring the discussion to a close by testing the limits of such differentiation, doing so by turning to Gilles Deleuze's, Michel Foucault's and Giorgio Agamben's understanding of the relationship between life, death and the possibility of freedom.

First, Deleuze's discussion of Joë Bousquet's confrontation with a life of pain suggests that there is an alternative to the investment in securing control of the time and manner of a loved one's or one's own death and the juridico-political effects of such investment. As observed in the previous chapter,

Bousquet rejected suicide and preferred to take charge of his life by accepting and embracing his pain. Bousquet wrote: 'To my inclination for death ... which was a failure of the will, I will substitute a longing for death which would be the apotheosis of the will ... Become the man of your misfortunes, learn to embody their perfection and brilliance.'[52] All notions of desubjectivation and depersonalization entail a trial of strength between opposing forces and a test of the self, which, as Foucault argues, is the course of subjectivation that Christianity bequeaths upon 'Western thought'. The public celebrations of Pretty's death with the words 'she is free at last' would seem to attest to such understanding. The true art of self, as Deleuze's evocation of Bousquet's words would seem to suggest, lies instead in openness to the fateful and unbridgeable process of dissolution, beyond all ritual and processes of resignification.

Secondly, it is unclear that Foucault and Agamben would disagree with this alternative understanding of autonomy and integrity, even authenticity, that Deleuze's evocation of Bousquet and his moving reflections on life and death would seem to advance.

Toward the end of his life, Foucault came to regard the confrontation with death as critically important to the formation of the subject. Foucault brings his lectures on *The Hermeneutics of the Subject* (1982) to a close by insisting that the construction of a form of existence beyond the demands of power will involve overcoming the understanding of life as a test of the self and that this will also involve a different comportment toward death. The final lecture in the series thus opens with a discussion of Classical reflections on the confrontation with death, something that, Foucault argues, was critically important to the constitution of the virtuous citizen. As he puts it, the meditation on death was so important because:

> Death is ... not just a possible event; it is a necessary event. It is not just an event of some gravity; for man it has absolute gravity. And finally, as we all know, death may occur at any time, at any moment.[53]

Foucault then refers to Stoic instructions about how to organize daily conduct so at to draw greatest advantage from such unpredictability. Living each day as if it were the last, the virtuous citizen should set out a plan of action each morning and then review the actions taken in the evening, before sleep. In so doing, these exercises aid the construction of truthfulness to oneself, a principle which, as Foucault observed, would eventually become the mainstay of the Christian examination of one's conscience. The critical point is, however, that these Stoic instructions were not orientated toward the renunciation of self to secure future salvation, as is the case with the Christian examination, but were orientated toward the past, aiming to ensure the best correspondence possible between knowledge and actions undertaken in the light of such knowledge. It is this correspondence that allows Marcus Aurelius to go to sleep fully satisfied, in the knowledge that he has 'lived'.[54] Such understanding is fundamentally important to Foucault's closing remarks that the true heir to this construction of the relationship between life, death and truth is nothing less than the ongoing project of enlightenment.[55] While the consequent normative injunction is to adopt a critical, but not denunciatory comportment toward the present, it remains unclear what might be a historically adequate, corresponding bodily practice, especially because there can be no going from an irretrievably Judeo-Christian culture backwards, to the Stoics and their world.[56]

The answer to the difficulty confronting Foucauldian thought is perhaps to be found in Agamben's writings. Agamben argues in his many reflections on the nature of biopolitical governmentality that this form of governance rests on the production of the subject by splitting a primordial state into human and animal and then mobilizing the relationship between the two that is thus instituted. He summarizes the process in *The Open*, writing of an 'anthropological machine [which] articulates nature and man in order to produce the human through the suspension and capture of the inhuman', the inhuman being here an alternative term for bare life.[57] On the same occasion, Agamben refers to Walter Benjamin's reflections on the dialectic of nature and culture, in which Benjamin sought to articulate that situation when humanity brings the dialectic to a halt, when it neither masters nature nor is it mastered by nature, but occupies a place or moment that is 'in between' the two terms. Agamben then seeks to characterize the figure that would emerge in the moment when this 'anthropological machine', pitting human and animal in relation to another, comes to a 'standstill'. Following Benjamin's lead, he suggests that sexual union can be understood as the moment when the differentiation operated by the machine and all that flows from its work is abrogated. Referring to Titian's 'The Nymph and the Shepherd', where, according to Agamben, Titian revisits themes first broached in 'The Three Ages of Man', Agamben writes that in the later painting:

> The enigma of sexual relationship between the man and the woman ... receives a new and more mature formulation. Sensual pleasure and love ... do not prefigure only death and sin. To be sure, in their fulfilment the lovers learn something of each other that they should not have known – they have lost their mystery – and yet have not become any less impenetrable. But in this mutual disenchantment from their secret, they enter, just as in Benjamin's aphorism, a new and more blessed life.[58]

In *The Ticklish Subject* (1999), Slavoj Žižek argues that, if this domain of bodies and pleasures is to be at all meaningful, it can only be so insofar as it relates to some other, elusive order, the order of fantasy and desire. The formative, endless struggle with desire always already unfulfilled can only be overcome by exiting the symbolic order altogether and becoming nothing. Žižek names this state of absolute evacuation 'symbolic death'.[59] This state on the boundary between sex and death is, however, precisely that which Agamben tries to capture in *The Open*, because true freedom lies in letting go of all investment in another place and another time. This passive, but affirmative, notion not only evokes Deleuze and Guattari's notion of erotic dissipation within the body without organs, but also seems not so distant from Joë Bousquet's affirmative embrace of death: 'To my inclination for death ... which was a failure of the will, I will substitute a longing for death which would be the apotheosis of the will ... Become the man of your misfortunes, learn to embody their perfection and brilliance.'[60]

It is not clear how the Stoicism that Deleuze, Foucault and Agamben would thus seem to evoke might be operationalized today, except perhaps by embracing the care offered by palliative medicine and the hospice movement. In the meantime, however, Foucault, Deleuze and Agamben offer constructions of the relationship between life and death that are very different from one another, Deleuze's being much more open to a construction of life no longer one's own, but the product of molecular processes that transcend the finitude of the mortal organism; and yet death would seem to matter equally to Agamben, Foucault and Deleuze. Just like the perspective advanced by the biogerontological project, Agamben, Foucault and Deleuze's is a thoroughly relational understanding of human existence, but one in which, at the same time and in the very same breath, the importance of mortality remains paramount. True to Friedrich Nietzsche, the shared challenge is to come to terms with finitude and in a way that acknowledges such fate in its fullest sense, refusing all possibility of salvation. As Nietzsche put it memorably:

> Whoever thou mayest be, beloved stranger, whom I meet here for the first time, avail thyself of this happy hour and of the stillness around us, and above us, and let me tell thee something of the thought which has suddenly risen before me like a star which would fain shed down its rays upon thee and every one, as befits the nature of light. – Fellow man! Your whole life, like a sandglass, will always be reversed and will ever run out again, – a long minute of time will elapse until all those conditions out of which you were evolved return in the wheel of the cosmic process. And then you will find every pain and every pleasure, every friend and every enemy, every hope and every error, every blade of grass and every ray of sunshine once more, and the whole fabric of things which make up your life. This ring in which you are but a grain will glitter afresh forever. And in every one of these cycles of human life there will be one hour where, for the first time one man, and then many, will perceive the mighty thought of the eternal recurrence of all things: and for mankind this is always the hour of Noon.[61]

How these words, whose meaning is inseparable from the awareness of mortality, are to be interpreted and enacted is, of course, indeterminate and their meaning depends on the historical context in which they are situated and repeated.[62] Such repetition would seem to be a necessary first step toward the 'decolonization' of death, toward the disconnection of life, death and the Judeo-Christian heritage.

Conclusion

As the archaeology of our thought easily shows, man is an invention of recent date. And one perhaps nearing its end. If those arrangements [of knowledge sustaining this invention] were to disappear as they appeared, in some event of which we can at the moment do no more than sense the possibility – without knowing either what its form will be or what it promises – were to cause them to crumble, as the ground of Classical thought did, at the end of the eighteenth century, then one can certainly wager that man would be erased, like a face drawn in sand at the edge of the sea.

MICHEL FOUCAULT

To have light and clarity alone, one would have to inhabit the very source of light, or, I don't know, remove the medium and thus create a void. As soon as the medium intervenes, the ray of light energetically and haphazardly seeks its fortune. One sees because one sees badly. It works because it works badly.

MICHEL SERRES

Coming back to Miranda's wistful last words by way of Crosby Beach, Antony Gormley describes his installation, which was originally intended for placement along the Elbe rather than the Mersey, in the following terms:

> The idea was to test time and tide, stillness and movement, and

somehow engage with the daily life of the beach. This was no exercise in romantic escapism. The estuary of the Elbe can take up to 500 ships a day and the horizon was often busy with large container ships.[1]

There is something deeply resonant about such play of contemplative stillness and the movement of transformation on the shores of something very much like Michel Foucault's beach. Embracing, but also seeking to keep the two moods apart, however impossibly, the past few chapters of *Biopolitics and the Philosophy of Death* have sought to articulate the contours of a genealogical answer to the question posed at the outset about how best to make sense of the present moment. The central thesis has been that, to answer this question, one needs to think about death, and not just as a philosophical concept, but also as an objective phenomenon of increasing interest to the biomedical sciences. The scale and scope of contemporary investment in the powers of medicine and biology to deliver longer and healthier lives is without precedent, every day assuming greater importance in the shaping of our daily lives, from the moment of our birth to that of one's death. In their different ways, Michel Foucault and Gilles Deleuze, yet unparalleled heirs to Friedrich Nietzsche's conjunction of philosophical and biological thought, provide a powerful set of tools with which to understand the relationship between life and death as it unfolds today, but the answers thus provided are far from decisive. Death seems to be unlike any other object of critical reflection, raising complexly knotted questions about both knowledge and method. *Biopolitics and the Philosophy of Death* has sought to illuminate how such complexity matters to any understanding of the present moment.

History and the end of Man

In the course of critical reflection upon Michel Foucault's understanding of the transformative processes involved in the constitution of the modern subject, Giorgio Agamben comes to regard Jacob Taubes as a fundamentally important thinker. Taubes opens *Occidental Eschatology* (1947), a series of reflections on the enduring importance of theological conceptions of human existence, by announcing pointedly that 'the subject of inquiry is the essence of history'. He then goes on to write that:

> The course of history is borne away in time ... The nature of time is summed up by its irreversible unidirectionality. From a geometrical point of view, time runs in a straight line in one direction. The direction of this straight line is irreversible.[2]

The meaning of these words is inseparable from the end of days. As observed earlier, the monotheism which Judaic and Christian theologies share posits the unity and meaningfulness of the diversity and multiplicity characterizing the world of all things apparent. Such unity and meaning are not self-evident, however. The community of believers and their God must then be alienated from one another, and the realization of their unity must also take the form of a revelation deferred. Furthermore, such revelation must disclose the falsity of the phenomenal world, and, as such, the realization of the truth revealed requires the overturning of the existing order of things. Judeo-Christian theology, in other words, cannot but be eschatological and apocalyptic. The enduring importance of this onto-theological framework, Taubes argues, lies in the conceptual transformations consequent upon the Copernican decentring of the *cosmos*. Theological thought turned inward, metaphysics was displaced by historical process, and the historical process itself became the vehicle of spiritual redemption. Taubes thus brings *Occidental Eschatology* to a close by writing:

> With Hegel on the one hand and Marx and Kierkegaard on the other, this study is not simply closed but essentially resolved. For the entire span of Western existence is inscribed in the conflict between the higher (Hegel) and the lower (Marx and Kierkegaard) realms, in the rift between inside (Kierkegaard) and outside (Marx).[3]

In other words, historical development is a matter of human struggles to break free of all necessity. Arguably, this is why history matters, and there is some splendour to the vision, but Taubes's argument also rests on the sharpest division between nature and culture. Taubes seems completely deaf to the like of Hans Jonas's considerations about the transformative impact of Charles

Darwin's *On the Origin of Species* (1859).[4] It is unclear therefore how Taubes would account for premonitions that the shattering of boundaries between nature and culture, between that which is given and that which is made, might lend considerable credibility to the contemporary notion that, thanks to the contemporary development of the biomedical sciences, 'the first person to live to be a thousand years is certainly alive today'. Agamben and Rosi Braidotti, figures greatly indebted to Foucault and Gilles Deleuze, offer two very different perspectives on such a situation.

In *The Open* (2002), Agamben suggests that the Darwinian turn and its accompanying decentring of *anthropos* could be understood as bringing the historical dialectic of nature and culture to a 'standstill'.[5] To better understand the nature of this extraordinary moment, it useful to recall Foucault's disclosure that the modern subject was the product of an impossible discursive topology, and that this figure was destined to erasure, just 'like a face drawn in sand at the edge of the sea'. Much has been written about the exact nature of this figure, Deleuze himself suggesting that such epochal moment should be understood 'much less than the disappearance of living men, and much more than a change of concept: it is the advent of a new form that is neither God nor Man and which, it is hoped, will not prove worse than it previous forms'.[6] Such a reading, however, evacuates Foucault's thought of all the more radical onto-historical ambitions that have exercised innumerable critics ever since the publication of *The Order of Things* (1966). The figure disappearing must instead be understood in its fullest somatic density, a density increasingly subject to biomedical intervention and transformation, if not, as some would say, enhancement. One need not agree entirely with Agamben's analysis of the present situation, which conjures images of human existence suspended between life and death, subject to the untempered, sovereign powers of medico-juridical reason, but he does appear to appreciate more than most the stakes involved when he writes that:

> If one day, according to a now-classic image, the 'face in the sand' that the sciences of man formed on the shore of our history should finally be erased, what will appear in its place … [will] … not represent a new declension of the man–animal relation so much as a figure of the 'great ignorance' which lets

both of them be outside of being, saved precisely in their being unsavable.⁷

This insight seems important to a critical evaluation of those arguments that would regard the current situation as heralding new opportunities for human emancipation, new opportunities to free the human animal from all forms of subjection, including the ultimate manifestation of such subjection, its subjection to the arrow of time. As observed in an earlier chapter, Braidotti dismisses all concern for human finitude:

> From the position of an embodied and embedded female subject I find the metaphysics of finitude to be a myopic way of putting the question of the limits of what we call 'life' … Death is overrated. The ultimate subtraction is after all only another phase in a generative process. Too bad that the relentless generative powers of death require the suppression of that which is the nearest and dearest to me, namely myself, my own vital being-there. For the narcissistic human subject, as psychoanalysis teaches us, it is unthinkable that 'life' should go on without my being there. The process of confronting the thinkability of a 'life' that may not have 'me' or any 'human' at the centre is actually a sobering and instructive process.⁸

Recalling Titian's 'The Three Ages of Man' and all the ambiguities of contemporary discussion of the meaning of life, it is not clear how such aspirations to the renewal of hope and community, predicated as each are on a notion life that transcends all that is situated and singular, marks anything but the promise of life eternal, and thus renews the onto-theological machine, today in the guise of Life Itself.

Seeking to eschew all such promises of redemption, *Biopolitics and the Philosophy of Death* has sought to explore the possibility of a critical comportment that is agnostic about the epochal nature of the present moment, and that seeks instead to test the limits of all existing critical apparatus, to borrow from Paul Rabinow's *Anthropos Today* (2005).

From multiplicity to heterogeneity

Biopolitics and the Philosophy of Death has sought to make sense of the present moment by focusing on modern biology in all its detail because, in many ways, biology has today taken on the burden once shouldered by theology and humanism to speak about the human condition, and focusing in particular on the point of conjunction between modern biology and philosophical reflection upon the meaning of human existence, namely ageing and death. Furthermore, precisely because the understanding of history that Jacob Taubes advanced in *Occidental Eschatology* can no longer be taken for granted because Man and God are both dead, the approach has been genealogical. The genealogy of the contemporary, attempted reconfiguration of biomedical understanding of ageing which biogerontology has sought to advance suggests that, once conceived as more than a matter of conceptual abstraction, there can be no overarching understanding of biological existence capable of encompassing seamlessly phenomena at the level of molecules, organisms and populations. However uneasily, the anatomo-political, biopolitical and molecular ordering of knowledge and power are today coexistent.

The coexistence of different constellations of objects, modes of conceptual organization and supportive institutional structures, toward which the notions of anatomo-political, biopolitical and molecular ordering of knowledge and power point, can be understood in the same terms as John Pickstone offers in *Ways of Knowing* (2001).[9] Briefly, Pickstone identifies three modes of producing knowledge. These are the practice of natural history, which depends on describing, collecting and classifying. Analytical practices are instead oriented toward discovering how things work by measuring and modelling relations between component parts. Experimentation, finally, seeks to produce knowledge by manipulating the component parts under controlled conditions. While such different ways of knowing have often been aligned diachronically, as is clearly the case in Michel Foucault's *The Order of Things*, Pickstone regards them as operating in parallel with one another, or better, as accumulating historically, so that, when it comes to the question of biomedicine, Pickstone might be regarded as offering a syncretic model, where others have proposed the convergence and synthesis of the different modes.[10] This said, if *Ways of Knowing* can then be regarded as a critical response to *The Order of Things*, Pickstone, like Michel Foucault before him, offers no account of what subjectivity and form of embodied existence might correspond to three of the ways of knowing and their syncretic combination.[11]

Nikolas Rose and Paul Rabinow, driven by the same concerns about questions unanswered, ask what forms of subjectivity corresponds to the contemporary emergence of the biomedical sciences as the pre-eminent technology of governance. As has been pointed out, they propose that contemporary subjects are constituted in the work of caring for their bodies, bodies that are understood as plastic assemblages of molecular components, in their exercise of biological citizenship and active participation in the evolution of a 'politics of life itself'. As Rose puts it toward the end of *The Politics of Life Itself* (2007), 'monitoring, managing, and modulating our capacities … is the life's work of the contemporary biological citizen'.[12] Yet, if the proposals advanced by the transatlantic, biogerontological coalition, on which the second part of *Biopolitics and the Philosophy of Death* has focused, are best suited to confront the challenges presented by an ageing population, and if these proposals can be regarded as instantiating those developments that have led Rabinow and Rose to call for this new understanding of political subjectivity, the fate of these same proposals seems to raise a number of questions about what might be regarded as the new, biosocial way of life. One is left with the sense that, when it comes to ageing and death, there is no new subjectivity, except perhaps that of a fractured subject, torn between the competing demands issuing from the anatomo-political, biopolitical and molecular ordering of knowledge and power. As Stephen Katz and Barbara Marshall have commented about ageing today, 'older individuals must cope with the impossible burden of growing older without aging'.[13] This impossible situation, arguably, must be the response to Peter Keating and Alberto Cambrosio's observation that:

> Some, taking their cue from historians such as Michel Foucault, have described the dispersion of 'the clinical gaze' through a fragmentation of the patient's body, whereby the latter 'is no longer localized in the discrete, integral body of the actual patient', but, rather, is distributed (figuratively speaking) among

a number of different specialties, and (in a literal sense) simultaneously present as a set of samples in different sections of a hospital … This fragmentation of the body is, moreover, held to mimic social fragmentation, as instantiated in the increasingly complex division of labor. Yet, to speak of a fragmentation of the patient is to lose sight of all the work that goes into keeping everything together, into making sure that, for instance, the sample that has left the body will rejoin it in terms of meaningful (for the task-at-hand) results.[14]

The call to synthesis and clarity is powerful, and the past few chapters have sought to respond otherwise to W. B. Yeats's melancholic words: 'things fall apart; the centre cannot hold; mere anarchy is loosed upon the world.'[15] In other words, *Biopolitics and the Philosophy of Death* has sought to resist all post-humanist temptation to find a new and alternative plane of consistency, and to maintain instead all things in their wholly unsettling dispersion, just watching and listening to the noise of pebbles on the beach being washed by the tidal ebb and flow on Crosby Beach.

From Rabinow to Boltanski and Thévenot

Finally, the call to synthetic conclusion and closure is powerful, so let me, authorial first person, bring *Biopolitics and the Philosophy of Death* to close by proposing that Paul Rabinow, Luc Boltanski and Laurent Thévenot would appear to provide the wherewithal to begin thinking more systematically about a critical comportment toward the heterogeneity that is situated wholly ambivalently between the stillness of thought and the urge to immersion in the movement of historical transformation.

In *Marking Time* (2008), Paul Rabinow, the anthropologist of the contemporary who is invested in speaking about the present moment in a way that maintains its openness, resisting all urge to forge answers to questions about what the future holds in store, evokes the notion 'adjacency' as a mode of critical engagement with both actors and readers. This is a mode of juxtaposing one's multiple and divergent objects of study that does not deny the distance between the observer and the observed, but neither does it insist that the observation reported must therefore be untrue to the phenomena themselves. The mode of adjacency is a business of:

> Listening, observing, hearing, querying, sensing, reflecting, pondering, wondering, and writing at various times, during, before, and after performing these other actions. Its goal is identifying, understanding, and formulating something actual neither by directly identifying with it nor by making it exotic.[16]

While resisting both mimetic and diremptive, Gnostic urges, this mode is capable of producing critical accounts of the present moment because it is disinterested, in the sense that the interests at issue are different to those motivating the actors observed. It is effective insofar as the juxtapositions produced capture the readers' interest, drawing their attention to unexpected connections and the possibility of a different understanding of the world around them. Marcel Duchamp and Gerhardt Richter are the models. Rabinow seeks to clarify this alternative critical mode by drawing on Gilles Deleuze's distinctions between potential and virtual states:

> This anthropological practice is characterized by what might be called a mode of virtual untimeliness. Let me explain. The difference between a mode of potentiality and what Deleuze has called a mode of virtuality consists in the fact that potentiality actualizes a state, a quality, or a form that is already inherent or resident in the being, thing, or process under consideration. The mode of virtuality does not directly partake of this metaphysical world. It operates adjacent to it, moving along side potentialities and actualities so that these can be taken up and refracted on another form. In another mode. That is to say, the virtual as opposed to the potential is a mode replete with real things and processes but redirected, removed from their habitual courses. Traditionally, my diagnosis and prescription is for those working within a space of anthropological *Bildung* not to look within but to be constantly working out, so as to create better forms for the self, for others, and for things.[17]

There are good reasons to be wary of any uncritical construction of the virtual, because the more it insists upon the insignificance of the organism, in all its actuality and finitude, the more it is at risk of falling back into some form of philosophical idealism and its material counterparts, from the *homunculus* of early modern imagination to Richard Dawkins's 'selfish gene'. As the past few chapters of *Biopolitics and the Philosophy of Death* have sought to explain, the understanding of life at play in the notion of the virtual exists parasitically, differently to, but not detached from, the singularity of a particular life, and it is forged on the boundaries between this particular life and the 'movement of the infinite'.[18] This said, Rabinow's intention is not to import into anthropological practice some philosophical reflection upon transcendence or immanence, but to develop an alternative mode of representation that is capable of apprehending the openness of the present moment, or, in other words, an alternative mode that is capable of creating a space of stillness in which critical reflection is combined with care for the full complexity and heterogeneity of the phenomenal world, rather than with deductions from first principles of any kind. On this understanding, writing has nothing to do with signifying, but, like the poet's writing, it is about mapping and promoting new ways of seeing and thinking, even projecting thought into regions still to come, into Louis Aragon's 'beyond'. There is something deeply valuable about the consequent understanding of writing as an experiment designed to provoke questions, rather than produce synthetic answers. This has been the aim of *Biopolitics and the Philosophy of Death* while juxtaposing anatomo-political, biopolitical and molecular modes of ordering knowledge and power, drawing out all 'the accidents, the minute deviations – or conversely, the complete reversals – the errors, the false appraisals, and the faulty calculations that gave birth to those things that continue to exist and have value for us'.[19]

There remains, however, something unsettlingly subjective and aesthetic about such professions of openness to diversity, reuniting heterogeneity as they do in the work of art.[20] From this perspective, there is also something very important about the argument advanced in Boltanski and Thévenot's *On Justification* (1991). As was noted in an earlier chapter, Boltanski and Thévenot ask a very practical question about how any two persons are able to construct shared modes of life and are able to do so despite all the differences between them that one might observe. For interesting and germane reasons, Boltanski and Thévenot are unable to offer any precise answer to this long-standing question, but in the course of articulating the nature of the problem, they develop a usefully complex understanding of social action. Basically, they posit a number of alternative modes of existence, which they label by association with the names of an equal number of notable philosophers. These modes will be invoked by anybody intent on constructing a common world, and the process of justification involved in forging a compromise will produce as much uncertainty as clarification. In other words, the very process of clarification and enlightenment necessary to construct the bonds of sociality is the very source of instability, establishing and magnifying differences. As such, Boltanski and Thévenot share Deleuze's interest in heterogeneity and multiplicity, but not necessarily Deleuze's vitalist metaphysics. More importantly, Boltanski and Thévenot insist that there can be no external, privileged position from which to evaluate and adjudicate between different orders, and herein lies their importance to *Biopolitics and the Philosophy of Death*.[21] From their perspective, any appeal to the work of art as a principle of unification partakes in Augustine's 'inspired polity'. As Boltanski and Thévenot put it, when describing the attributes and values that characterize the 'inspired world':

> In this world, persons may be more or less worthy inasmuch as they are capable of experiencing the outpouring of inspiration and thus of acceding to perfection and happiness ...
>
> The passion that moves them instils in them ... a desire to create – a desire awakened by inspiration – along with anxiety or doubt, love for the object pursued, and suffering.
>
> The worthiest person in terms of inspiration are often despised by the world at large; they may be poor, dependent, and useless. But their deficient state actually enhances their access to knowledge of the world's truly harmonious figures (heaven, the imaginary, the unconscious and so on). This holds true for children, who are 'curious, inventive, passionate', for women, the simple-minded, and madmen, and also for poets, artists (who are of the same 'nature as women'), monsters, 'frightening, imaginary creatures', and other freaks of nature. In this world, in which beings are appreciated for their uniqueness and in which the being with the highest degree of generality is the most original, the worthy are at once unique and universal.

One shifts without transition from the singularity of the I to the generality of Man. Thus artists, who often embody inspired worth today, are worthy because they include the others in the singularity of the proper name: Baudelaire, Cocteau, Einstein, Galileo, Mozart, Shakespeare, and the like.[22]

God and Man are products of a particular manner of constructing the common good and the artist is today's inspired figure who is able to evoke and communicate the essential qualities of such foundational modes of recognition. Significantly, while the Nietzschean tones of Boltanski and Thévenot's summary of the inspired mode are inescapable, Friedrich Nietzsche provides no alternative and definitive understanding of the common good. As Boltanski and Thévenot observe, Nietzchean relativism escapes solipsism and nihilism by positing its own general principle of worth, power. The metaphysics of power, which Nietzsche espouses and provides the foundations for all Nietzschean critiques, including Deleuze, Foucault and all critical relativism, is no more than another of the different modes of constructing that which is in common, Thomas Hobbes's 'polity of fame'.[23] Seeking to abjure the purification and the clarity of univocity, one would then have to say that the contemporary subject is produced at the intersection of incommensurable and irreducibly different regimes, in the contemplation of the impossibility of any supervening, rational and analytical ordering of experience. This said, if God and Man can no longer provide any definitive compass, the same is true of Life Itself. As Boltanski and Thévenot also observe, there is something deeply valuable about the Nietzschean notion of the 'vitality of life itself', but it is not clear how such value is to be understood, once it is disconnected from the value of a particular life, in all its finitude, except at the cost of disqualification as an 'illegitimate' value because such disconnection must necessarily fall short of the assumption of common dignity that founds all proper polities.[24] The centre does not hold, but neither do things fall apart, nor is mere anarchy loosened upon the world. Instead, we are today called by others to invent new ways of being ourselves, flitting between certainty that one day we will be no more and uncertainty about how best to postpone that fateful day and live the days in between in the most meaningful and productive manner. Whether death comes to us or not, it is still remains with us, shaping our thinking about ourselves and others.[25] This is undoubtedly an uncertain state of affairs, but, as Michel Serres, whose words preface these closing reflections, might put it, this situation is a better place to start than from the blinding truths of first principles, be they God, Man or Life Itself.[26]

NOTES

Introduction

1 Basic demographic observations such as the quadrupling of centenarians' number over the last thirty years lend considerable credibility to de Grey's claims. For an introduction to de Grey's ideas and the importance some critics attach to these ideas, see Agar (2010): 83–106.

2 In *Death and Mortality in Contemporary Philosophy* (2005), Bernard Schumacher reviews very usefully the place of death in philosophical traditions to which Michel Foucault, Gilles Deleuze and others are indebted, but the approach taken here is perhaps closer to Benjamin Noys's *The Culture of Death* (2005), at least insofar as *Biopolitics and the Philosophy of Death* seeks to engage with developments outside the disciplinary boundaries of philosophical argument and explore their implications for the latter.

3 Gil Anidjar draws attention to the tension within contemporary, critical reflection upon the meaning of 'life', divided as it is between understanding of the term as referring to a historically specific discursive construct and understanding of the same term as referring instead to a phenomenon that transcends all the vagaries of human history.

4 For an extended discussion of this and another of Titian's paintings in relation to Giorgio Agamben's argument about human, embodied existence, see Palladino (2010); also Oliver (2007) and Palladino (2013).

5 Viewed from a strictly philosophical perspective, the argument advanced can be regarded as an exploration of the relations Giorgio Agamben charts in 'Absolute Immanence' (1996), but focusing particularly on death and human embodied existence; see Smith (2012): 271–86. Furthermore, *Biopolitics and the Philosophy of Death* builds on existing reflections upon philosophy, death and human embodied existence since the focus is on contemporary

biomedical constructions of ageing and death, constructions that do not always start from the organism and are not constrained by the privilege often accorded to the organism; cf. Schumacher (2005).

6 For an introduction to the importance of Gilles Deleuze and neo-Deleuzian thought to critical studies of science, technology and medicine, the field from which *Biopolitics and the Philosophy of Death* emerges, see Jensen and Rödje (2009).

7 Hans Jonas's *The Phenomenon of Life* (1966) still offers an unparalleled analysis of the relationship between biology and the secularization of theological preoccupations.

8 Jonas (2001): 53.

9 In *Being and Time* (1927), Martin Heidegger sought to distinguish between ontic and ontological orders. While the former referred to the descriptive characteristics of a particular thing and the basic facts of its existence, the latter concerned the understanding and investigation of Being, the ground of Being, and the concept of Being itself. As will become evident, distinction between the two orders is fraught and the embrace of difficulty is not easily reduced to unwarranted naturalism.

10 Foucault (1996): 408. The historical background to this shared, resolute anti-humanism is discussed very helpfully in Geroulanos (2010). For an equally helpful discussion of the related psychoanalytic configuration of the 'death of Man', which Michel Foucault and Gilles Deleuze also rejected, see Borch-Jacobsen (1991): 1–20.

11 Nietzsche (1990): 162.

12 Foucault (1977): 146.

13 Serres (1980): 128. On the proximity of Michel Serres and Jacques Derrida's thought, see Boyne (1998).

14 Foucault (1994b), 387.

15 For a discussion of the work involved in disconnecting embodied existence from all reference to the Judeo-Christian heritage as work of 'decolonization', see Anidjar (2011). For a discussion of the manner in which some of the developments discussed here might unfold in a very different social and cultural context, arguably far removed from the Judeo-Christian heritage, see instead Cohen (2000).

Chapter 1: Evental figures and questions of method

1. Kirkwood (1999): 243.
2. Kirkwood (1999): 256.
3. Attesting to the intensity of contemporary preoccupation about ageing and dying, *Time of Our Lives* positioned Thomas Kirkwood as a leading public figure, resulting in his invitation to deliver the prestigious BBC Reith Lectures, which have done so much over the years to extend understanding of science among the global public within the reach of the British Broadcasting Corporation, and to speak on 'The end of age' (2001).
4. Kirkwood (1999): 242.
5. See also Lakoff and Johnson (1980).
6. Donna Haraway's much overlooked *Crystals, Fabrics, and Fields* (1976) is still today a particularly important work because it offers a rigorous analysis of the work of metaphor within biology, usefully removed from Haraway's later, distinct preoccupation with the implications of biology for the development of post-humanist thought, which Haraway sets out in *Modest Witness* (1997). Evelyn Fox-Keller's nearly contemporaneous *Making Sense of Life* (2002) exemplifies equally well how the analysis of the work of metaphor within biology can be separated from preoccupations about the nature of the human subject. This issue will be examined further in a later chapter and with reference to John Protevi and James Griesemer's reflections on contemporary biology.
7. As Michel Foucault puts it on the first page of *The Birth of the Clinic*, 'for us, the human body defines, by natural right, the space and origin and distribution of disease: a space whose lines, volumes, surfaces, and routes are laid down, in accordance with a now familiar geometry, by the anatomical atlas. But this order of the solid, visible body is only one way – in all likelihood neither the first, nor the most fundamental – in which one spatializes disease. There have been, and will be, other distributions of illness'; Foucault (1994a): 1. Famously, Foucault claims that the current geometry emerges as a response to the 'impossible spatial synthesis' developed under the ontological theory of disease; Foucault (1994a): 15.
8. Admittedly, the phrase 'post-structuralist juridico-political theory' is awkward, but it is also necessary. First, *Biopolitics and the Philosophy of Death* emerges from concern to examine the social,

as opposed to cultural, determination of epistemic structures. Secondly, contrary to much of the literature on 'governmentality', the phrase 'post-structuralist juridico-political theory' does most justice to the complexity of Michel Foucault's arguments about the construction of the modern subject, in which neither social, political nor juridical modes of production is granted priority over the other. Such organizational multiplicity, as *Biopolitics and the Philosophy of Death* will seek to establish, matters. The key texts for this perspective on post-structuralist theory are Dean (1994, 1999, 2013). Furthermore, the insistence on a positive configuration is intended to point indirectly to Jacques Derrida and his own interventions within recent debates about biopolitical order. Derrida positioned himself in relation to these debates in the essays collected under the title *The Beast and the Sovereign* (2009, 2011). Furthermore, Roberto Esposito (2008) and Kelly Oliver (2013) can be said to bring Derridean deconstruction to bear on the debates about the nature of biopolitical governance. This said, Derrida was interested primarily in language and only secondarily in death, the latter as a figure for the incompleteness of any linguistic system and its productive effects. Derridean deconstruction extends the assimilation so that it becomes all-encompassing. Viewed from this perspective, Derrida and Derridean deconstruction seem less useful to understand the diversity and differentially specific nature of biomedical endeavours to offer a positive representation of death, but they are not wholly irrelevant; see, for example, Rheinberger (1997). The overall thrust of *Biopolitics and the Philosophy of Death* is that, while privileging a positive construction, positive and negative configurations are in fact inseparable and intertwined and what is at issue is the importance to be attached to the organism.

9   Foucault (1994b): 387. The notion of event mobilized here is situated ambiguously between Michel Foucault's seemingly epistemological considerations and Gilles Deleuze's more assuredly ontological understanding of the event; see Deleuze (1969) and Foucault (1980). The importance attributed to Miranda is also indebted to Paul Rabinow's reflections on method, in *French DNA* (1999), whereby an otherwise unnoticed conjunction serves powerfully to open up the complexity of the present biopolitical moment and its manifold possibilities. For an alternative approach to connecting literary and cinematographic representations of ageing and death, such as Anthony Trollope's *The Fixed Period* (1882) and David Fincher's cinematographic adaptation of F. Scott Fitzgerald's *The Curious Case of Benjamin Button* (1922), and their broader historical context, see the essays by Andrea Charise,

Marlene Goldman and Cynthia Port in a special issue of *Occasion*, on 'Aging, Old Age, Memory, Aesthetics' (2012).
10. For an introduction to the importance of *The Selfish Gene* in the history of neo-Darwinism and its sociobiological extensions, see Segerstråle (2000): 53–79.
11. Dawkins (1976): 35.
12. Hayles (2001): 150.
13. See Pepper and Herron (2008): 621.
14. Foucault (1991): 28.
15. See also Judith Butler's important attempt, in *Bodies that Matter* (1993), to forge a fully materialist understanding of the subject that might allow for the discursive construction of such materiality. Donna Haraway's most decisive intervention is not wholly sympathetic, but is nonetheless more open to Gilles Deleuze and the neo-Deleuzian answers discussed below; Haraway (1997). See also Dianne Currier's discussion of the ways in which neo-Deleuzian thought might help to resolve some of the difficulties which Haraway's post-humanist thought leaves unresolved; Currier (2003).
16. Hayles (2001): 158.
17. Žižek (1999): 1. Slavoj Žižek, heir to Jacques Lacan, has turned to psychoanalysis to advance a critique of Marxist conceptions of ideology and the notion that false consciousness prevents the subject from seeing how things really are. Žižek argues that the subject's deepest motives are unconscious and that ideology serves to construct a liveable reality, but that such construction is far from seamless. In *The Ticklish Subject*, Žižek argues further that the critique of all belief in some single power, which post-structuralist theory advances, rests on the tacit assumption of an alternative plane whereby the critic can escape the unbearable burden of freedom disclosed by the seams and *aporiae* within the same ideological fabric. Judith Butler is one of Žižek's principal targets; see Žižek (1999): 247–312.
18. Braidotti (2007). Rosi Braidotti's essay on 'Biopower and Necropolitics' is a summary of the arguments advanced in *Transpositions* (2006). Braidotti's undertanding of psychoanalytical structures is indebted to Gilles Deleuze, rather than to Sigmund Freud and Jacques Lacan. On Deleuze and psychoanalysis, see Adkins (2007).
19. See Simmel (2007). For a very useful and illuminating discussion of Georg Simmel's understanding of life and death, see Pyyhtinen (2012).

20 Hayles (2001): 158. For an introduction to Rosi Braidotti's contribution to the renewal of materialism, see Dolphijn and van der Tuin (2012). The evolution of Rosi Braidotti's materialism is marked out by the trajectory from *Nomadic Subjects* (1994), through *Metamorphoses* (2002), to *Transpositions* (2006). Katherine Hayles's alternative understanding of post-humanism is marked out instead by the trajectory from *How We Became Posthuman* (1999), through *My Mother was a Computer* (2005), to *How We Think* (2013). In light of Hayles's criticism of Gilles Deleuze, figure just as important to the evolution of the 'new materialism', Arthur Kroker's confidence that Hayles is readily aligned with this line of thought seems misplaced, but Kroker's discussion remains very useful nonetheless; see Kroker (2011): 63–99. For a more general introduction to the new materialism, see Bryant, Srnicek and Harman (2011).

21 See also Braidotti (2006): 36–42.

22 As will be discussed in later chapters, Martin Heidegger's reading of August Weismann was indirect, but it was nonetheless formative. Briefly, as William McNeill (1999) and Thomas Kessel (2011) have observed, contemporary arguments about the relationship between the multicellular organism and its component parts were critically important to the evolution of Heidegger's understanding of human embodied existence. Heidegger does not appear to have ever referred to Weismann himself, but Weismann was deeply involved in these same debates and Eugen Korschelt, whom Heidegger does cite, brought Weismann's views to bear on the relationship between multicellular organization, ageing and death, so redoubling the importance of biological debates to the evolution of Heidegger's understanding of the subject; see also Krell (1992).

23 Mayr (1982): 698.

24 Dawkins (1976): 12.

25 See Sauvagnargues (2013); also May (2005) and Ansell Pearson (1999).

26 See Moore (2002); also Richards (2002). As will become evident in later chapters, biology refers here to a form of thought that only emerged toward the end of the eighteenth century.

27 Nietzsche (1983): 67; see also Emden (2008). Significantly, as Zygmunt Bauman observes in *Mortality, Immortality & Other Life Strategies* (1992), the evolution of historical consciousness must be viewed as a response to the finitude of human existence.

28 Foucault (1977): 146.

29 Foucault (1977): 153.
30 See Agamben (2009): 110–11.
31 See Foucault (2011); also Flynn (2005): 260–82.
32 From *The Time that Remains* (2000) onward, Jacob Taubes has acquired increasing importance to Giorgio Agamben's attempt to synthesize onto-theological and governmental understanding of power, which Agamben first began to forge in *Homo Sacer* (1995). For a critical introduction to Taubes's work and thought, see Assmann, Assmann and Hartwich (2010).
33 Taubes (2009): 3–4.
34 Taubes (2009): 191.
35 See Rabinow (1996c). Jacob Taubes's position in the debates to which Hans Blumenberg was contributing is evident in Taubes's 'Notes on Surrealism' (1966).
36 The redemptive, if not messianic, ambitions of Michel Foucault's historiography are nowhere more evident than in *History of Sexuality* (1976).
37 Taubes (2009): 5; see also 11–15.
38 Taubes (2009): 40.
39 Taubes (2009): 4.
40 Arendt, as quoted in Anidjar (2011): 715.
41 Taubes (2010): 263–4.
42 Taubes (2010): 264.
43 Jacob Taubes's discussions in *Occidental Eschatology*, wherein 'nature' comes to be equated with the politics of 'blood and soil', leave little room to doubt that the reference is to national socialism and its racial policies; see Taubes (2010: 40) and Assmann, Assmann and Hartwich (2010: xlv–xlvi). For a useful discussion of the relationship between Arnold Gehlen's thought and national socialism, see Magerski (2012). For a more general discussion of Gehlen's contributions to sociological understanding of embodiment, see instead B. S. Turner (1992) and Esposito (2011).
44 Foucault (2011): 180.
45 Foucault (1977): 164.
46 Recalling Jacob Taubes's criticism of Arnold Gehlen and the complicity of his assumptions with murderous destruction, Foucault notes, almost in passing, that 'in contemporary German philosophy since the war you can … find a whole problematization of Cynicism in its ancient and modern forms. And it is undoubtedly something to

be studied more closely: why and in what terms has contemporary German philosophy posed this problem?'; Foucault (2011): 179. Foucault leaves the question unanswered. This is perhaps yet another of those conjunctions where, as Giorgio Agamben charges, Foucault fails to inquire into the full implications of the historical situation wherein denuded existence comes to be regarded as the 'site of emergence of the truth'; see Agamben (1998): 6.

47 The term 'entanglement' adopted here is indebted not just to Michel Callon's distinctive understanding of interaction and association, but also to Michel Serres' formative understanding of debt; see Brown (2002).

48 For a discussion of the renewal of the pragmatist tradition, see the special issue on the subject which appeared in the *European Journal of Social Theory*, particularly Paul Blokker's very useful introduction; Blokker (2011). For a discussion of parallel developments within the field of science and technology studies, see instead Guggenheim and Potthast (2012).

49 Boltanski and Thévenot (2006): 1.

50 Boltanski and Thévenot (2006): 336.

Chapter 2: Ageing and the molecular way of life

1   Medawar and Medawar (1977): 159.
2   Foucault (1991): 138.
3   Foucault (1991): 139.
4   Foucault (2004): 251–2.
5   For a concise summary of the debates over the relationship between the disciplines involved in the articulation of these two governmental formations, see Colgrove (2002).
6   See also Seale (1998).
7   Alliance for Aging Research (2005): 1.
8   See Katz and Marshall (2003, 2004).
9   See Fries (2005). On the history of the Milbank Memorial Fund, see instead Fox (2006).
10 For a critical review of the debates over old age and assisted death, see Moody and Sasser (2012); also Moody (2001). Significantly,

while focusing on technologies of capital punishment, Kelly Oliver asks very helpfully whether the quest for control over the timing and manner of death might involve a disavowal of death and its sovereignty; Oliver (2013): 176. Viewed from this perspective, the distance between the quest for immortality and the insistence on a good death is less than would appear upon first consideration.

11 Alliance for Aging Research (2005): 4.
12 House of Lords (2005): 7.
13 House of Lords (2005): 103.
14 De Grey (2005a). The success of TED lecture as a vehicle to deliver innovative ideas has created much consternation within the scientific community, but it would seem that it has yet to be subjected to any systematic, critical scrutiny; see Robbins (2012). See also Agar (2010).
15 SENS Research Foundation (2013).
16 See Cooper (2006, 2008a).
17 On Michael West, see Hall (2003): 60–77.
18 Mass Observation Research Institute (2006): 14.
19 House of Lords (2005): 8.
20 Kirkwood (2005). Following the editorial refusal to allow de Grey any opportunity to respond, the review precipitated an excited debate in the British media. See de Grey (2005b); also Morelle (2006).
21 Mass Observation Research Institute (2006).
22 Alliance for Aging Research (2005): 12.
23 Butler, Miller et al. (2008).
24 See R. N. Butler (1975, 2008).
25 See Chadarevian and Kamminga (1998); Fujimura (1996).
26 Keating and Cambrosio (2003): 69–77.
27 Bernard, as quoted in Canguilhem (1991): 67.
28 Kaufman, Shim and Russ (2004): 737. On the polarized tenor of the public discussions following the publication of the President's Council on Bioethics report on medical enhancement, see Asch (2005); also Mahowald (2005) and Mykytyn (2008).
29 Kaufman, Shim and Russ (2004): 737.
30 Butler, Miller et al. (2008): 399.
31 Butler, Miller et al. (2008): 399.
32 See also Clarke et al. (2003). For a review of the literature on

biopolitical governmentality that focuses particularly on the contemporary transformation of embodied forms of life, see Lemke (2011): 93–116.

33 For a critical introduction to the conceptual complexity of the gene, see Fox-Keller (2000). The notion of the gene as an epistemic object is indebted instead to Rheinberger (1997).

34 Rabinow (1996a): 99. On Paul Rabinow's nominalism and his aversion to ontological speculation, see Rabinow (1988) and (1999b).

35 Rose (1999): 233; see also Rose (1985).

36 Rose (1990); 262; see also Rose (2007): 41–76.

37 Rose (2007): 223.

38 Rose (2007): 49.

39 Dawkins (1976); 35.

40 Holliday (2007): 105.

41 Holliday (2007): 106.

42 Butler, Miller et al. (2008): 399.

43 For an earlier attempt to articulate this proximity between biographical approaches and related concerns about the processes of subjectivation, see Palladino (2001).

44 The present enterprise might then be aligned most readily with Stephen Katz's *Disciplining Old Age* (1996), an outstanding critical reading of the major texts and institutional formations of modern gerontology, which inverts the historical approach taken by the like of Andrew Achenbaum and Hyung Wook Park, seeking to illustrate the ways in which gerontology, as a discipline, constructed its subject matter, the ageing body. At the same time, while Katz orders the diversity and complexity of gerontological thought very impressively, his archaeological approach forces a troublesome detachment and distance from the knot that arguably lies at the very heart of the difficulties confronting not just gerontology, driving a distinctively multidisciplinary approach to the construction of its object, but that also trouble modern thought itself.

Chapter 3: The evolutionary biology of ageing and death

1 Dawkins (1976): 12.

2   Ansell Pearson (1999): 189–99.
3   Deleuze (1990): 359–70.
4   See Winther (2001); also Churchill (1987); Griesemer and Wimsatt (1989).
5   See also Griesemer (2005).
6   See Sarkar (1992).
7   Mayr (1959): 2; see also Provine (2004).
8   See Haldane (1941).
9   See Medvedev (2000).
10 See Deleuze (1990): 362.
11 Comfort (1956): vii.
12 Comfort (1956): 1–2.
13 Comfort (1956): 9.
14 Weismann, as quoted in Comfort (1956): 9.
15 Weismann (1891b): 24–5.
16 Comfort (1956): 9.
17 Comfort (1956): 46.
18 Comfort (1956): 39. Though always interested in philosophy, Alex Comfort does not refer to Martin Heidegger until much later in life, in *Reality & Empathy* (1984).
19 Comfort (1956): 40–1.
20 See also Comfort (1950).
21 See Thane (2000): 344–5; also Harper and Thane (1989).
22 See Yapp and Bourne (1957).
23 Wolstenholme and Cameron (1955): 242.
24 Wolstenholme and Cameron (1955): 243.
25 Wolstenholme and Cameron (1955): 243–4.
26 For a useful discussion of the notion of 'régime', see also Sherratt (2006): 153–63.
27 Foucault (1980): 124–5.
28 See Gaudillière (2002); also Keating and Cambrosio (2003): 49–82.
29 See also Haber and Gratton (1994): 165–8.
30 Shock (1947): 94.
31 Shock (1961): 1026.
32 See Thane (2000): 436–57.

33 Le Gros Clark and Pirie (1957): xiv.
34 Le Gros Clark (1957): 90. On Frederick Le Gros Clark and ageing, see Thane (2000): 393–8.
35 Martin (1995): 458.
36 See Katz (1996): 73–6.
37 Sheldon (1948): 8; see also Thane (2000): 408–11.
38 The support which the Medical Research Council provided for research that has come to be regarded as fundamentally important to the emergence of contemporary understanding of Alzheimer's Disease is wholly consistent with this analysis, since the primary motivation of such research was to establish the physiological basis of psychological states and the age of the subjects examined was regarded as secondary; see Wilson (2014).
39 Katz (1996): 161–2.
40 See Stewart (2008); also Keating and Cambrosio (2003): 25–47.
41 Amulree, as quoted in Martin (1995): 460.
42 On social medicine and the uncertain boundaries between medicine and social science, see D. Porter (1997a).
43 Wolstenholme and Cameron (1955): 242.
44 Comfort (1957): 23.
45 See also Alberto Cambrosio, Peter Keating, Thomas Schlich and George Weisz's introduction to a special issue of *Social Studies of Science* dedicated to role of standards and regulatory objectivity within the biomedical sciences; Cambrosio, Keating, Schlich and Weisz (2009).
46 Treas (2009): 82.
47 See also Thévenot (2009).
48 On diseases of civilization, see R. Porter (1993).
49 On the precarious status of social medicine, see D. Porter (1997b).
50 Medvedev (2000): 898.
51 See Armstrong (1983): 85–92.
52 See Keating and Cambrosio (2003): 49–82.
53 Medawar and Medawar (1977): 159.
54 Comfort, as quoted in Goff (2009): 71.
55 Katz and Marshall (2005): 5; see also Katz and Marshall (2004).
56 Comfort (1979): 12.
57 For an interesting, near-contemporary response to Alex Comfort's

reflections on religion, which draws attention to Comfort's contribution to the counterculture, see John C. Poynton's essay on 'Alex Comfort on Self and Religion' (1980). On the links between the counterculture and technological innovation, see Turner (2006); Vettel (2008). See also Cooper (2008a): 15–50.
58   University College London (2013a); see also Patridge (2010).
59   University College London (2013b).

Chapter 4: Molecularizing the biology of ageing and death

1   See Friedman (2008).
2   Hall (2003): 14–41.
3   On Georges Canguilhem and the importance of his thinking about biology and medicine to the evolution of contemporary philosophy, see Geroulanos (2009); see also Rabinow (1994).
4   Bernard, as quoted in Canguilhem (1991): 67.
5   Canguilhem (1991): 76.
6   Keating and Cambrosio (2003): 20.
7   Keating and Cambrosio (2003): 76–77.
8   Significantly, Peter Keating and Alberto Cambrosio cite Paul Atkinson's *Medical Talk and Medical Work* (1995) as exemplifying the blindness to the biomedical work of integration to which they refer, but Anne-Marie Mol's *The Body Multiple* (2003) could be said to illustrate just as well the more general claims to which Keating and Cambrosio wish to respond. From this perspective, the attention to material practices is no answer because it can prove just as centrifugal as any exclusive preoccupation with concepts and discourses. The issues involved are addressed in important sociological interventions such as Zygmunt Bauman's *Life in Fragments* (1995) and Bryan Turner's *Body and Society* (1996).
9   Hayflick (1974): 37, 38.
10 Hayflick (1974): 38.
11 Hayflick (1974): 38.
12 Hayflick (1974): 39.
13 Hayflick (1974): 40.
14 Hayflick (1974): 40.

15 Hayflick (1974): 40.
16 Hayflick (1974): 43.
17 For a wider ranging analysis, more attuned to the cultural context and implications of the process of biological standardization, see Rader (2004); also Griesemer and Gerson (2006).
18 See also Amsterdamska (2005); Susser (1985).
19 See Lindee (2005): 90–119.
20 On Peyton Rous and the experimental system supporting viral oncogenesis, see Gaudillière (2002): 167–84.
21 Hayflick (1965): 614.
22 Hayflick (1965): 627 and 629.
23 Hayflick (1965): 629.
24 On developments at Stanford University, see Fujimura (1996): 37–53.
25 See Bookchin and Schumacher (2005): 118–31.
26 Lockett (1983): 59–63.
27 As quoted in Moreira and Palladino (2011): 321.
28 Eisenhower, as quoted in Moreira and Palladino (2011): 320.
29 As quoted in Moreira and Palladino (2011): 320.
30 As quoted in Moreira and Palladino (2011): 320.
31 Fleming, as quoted in Moreira and Palladino (2011): 320.
32 Shock (1961): 1026.
33 Wolstenholme and Cameron (1955): 244.
34 Hayflick and Moorehead (1961): 614.
35 Hayflick (1965): 633.
36 Shock (1966): vii–viii.
37 Maynard Smith (1962): 119.
38 See Hay and Strehler (1967); Strehler (1967).
39 Lockett (1983): 85.
40 Achenbaum (1995): 219–49.
41 Achenbaum (1995): 203–6.
42 Achenbaum (1995): 122–3.
43 Strehler (1967): 662.
44 Hayflick (1968): 37.
45 On Bernard Strehler and his increasing interest in August Weismann, see Strehler (1969).

46 Comfort (1957): 22.
47 Hayflick (1974): 43.
48 Hayflick (1974): 43.
49 See also Ballenger (2006): 75–80.
50 See Hall (2003).
51 See Rabinow (1996b); also Wilson (2007).
52 R. N. Butler (1977): 8.
53 R. N. Butler (1975): 401.
54 See Ballenger (2006): 123 and 118–28.
55 Miller (2002): 167.
56 See Ballenger (2006): 146–51; also Lock (2013).
57 Foucault (1994): 144.
58 Cowdry (1939): 682. Edmund Cowdry's notion of a '*milieu interne*', which he sought to articulate on the occasion of the Woods Hole Conference on the Problems of Aging, was no more than a restatement of Claude Bernard's formative evocation, nearly eighty years earlier, of a physiological '*milieu intérieur*'. For a more general discussion of the organism and the history of questions of integration it has posed, see Cross and Albury (1987).
59 Canguilhem (1991): 65–89.
60 Canguilhem (1991): 275–87.

Chapter 5: Forging the future

1 Nietzsche (1983) and Foucault (1977); see also Palladino (2002).
2 Kirkwood and Cremer (1982): 108.
3 Moreira (2000): 425.
4 Thévenot (2009): 797; see also Moreira, May and Bond (2009).
5 Dawkins (1976): 42–3.
6 Dawkins (1976): 44–5.
7 See Nelkin and Lindee (1996): 53–4.
8 Kirkwood and Holliday (1975): 482.
9 Kirkwood and Holliday (1975): 495–6.
10 Foucault (2004): 246.
11 See Kirkwood (1977); Kirkwood and Holliday (1979).

12 Kirkwood and Holliday (1979): 435–6. Richard Dawkins's *The Selfish Gene* is best regarded as popularizing George Williams and William Hamilton's contributions to the mathematical analysis of evolutionary process, including behavioural ecology. On the more general importance of the meeting that Kirkwood and Holliday were attending to the development of sociobiology and the surrounding debates, see Segerstråle (2000): 53–79.
13 See Dawkins (1976): 12.
14 Weismann (1891b): 145.
15 Wolstenholme and Cameron (1955): 242.
16 Foucault (2004, 2009); see also Bayatrizi (2008).
17 Foucault (2009): 10.
18 Foucault (2009): 77–8.
19 See Gigerenzer et al. (1989); also Hacking (1990).
20 Foucault (2004): 249–50.
21 Foucault (2004): 251–2.
22 Foucault refers to 'the theory of degeneracy'; Foucault (2004): 251–2. Historians of genetics have charted in considerable detail the great importance of degeneration to the emergence of genetics; see Paul (1995).
23 Kirkwood and Austad (2000): 237.
24 Kirkwood and Austad (2000): 235.
25 Kirkwood (1999): 242.
26 There is, as yet, no systematic historical study of the developments discussed in this section. The argument is based mostly on Katzman and Bick (2009) and Wilson (2014). See also Edwardson and Kirkwood (2002).
27 Kirkwood (2006): 81; see also Hughes (2006); Moreira, May and Bond (2009).
28 Olshansky, Hayflick and Carnes (2002): 292.
29 Butler, Miller et al. (2008): 399. See also the guide to Newcastle Campus for Ageing and Vitality, which informs the reader that: 'Today, the Newcastle Campus for Ageing and Vitality is a globally unique, large-scale development bringing together world-leading scientific and medical research with innovative health care, industry, civic agencies, and the public. The Campus, created through the Newcastle Biomedicine partnership between Newcastle University and the Newcastle Hospitals NHS Foundation Trust, aims to lead in finding new solutions to the many challenges and opportunities of

ageing populations. The University's Institute for Ageing and Health is based in the Academic Quarter where one of the major funders has been the Wellcome Trust. It features a series of unique facilities including: the Henry Wellcome Laboratories for Biogerontology; the Biomedical Research Centre and Biomedical Research Unit; and the Clinical Ageing Research Unit where early detection of age-related disease and intervention treatments is carried out. It also hosts the Newcastle Brain Tissue Resource. Business and retail quarters are also being developed on the Campus. Businesses are being helped to take advantage of the massive opportunities to provide older people – the largest growing section of the population – with the products and services they need to maintain independent, good-quality lives. Changing Age for Business was created by the University to provide access to our extensive expertise to help firms seize those opportunities. This includes incubator space for fifteen "age-related" businesses to work on projects alongside University researchers'; University of Newcastle (2014): 5. For a discussion of the transformation of healthcare which such initiatives illustrate, though tied to developments in the United States, see Clarke et al. (2003).

30   For a helpful discussion of such movement and consequent disruption of the existing alignment of ageing and American biomedical institutions, see Mykytyn (2006b). More importantly, the instability considered here is historical, being the product of expanding connections, rather than the essential instability toward which Oliver (2013) and Shershow (2010) and (2013), for example, point in their own discussions of contemporary constructions of dying and death.
31   House of Lords (2005): 7.
32   House of Lords (2006a): para. 9.
33   House of Lords (2006b): para. 6.
34   House of Lords (2006b): para. 6.
35   House of Lords (2006a): para. 9. 48.
36   Department for Work and Pensions n.d.
37   Weismann (1891c): 145.
38   Kirkwood and Austad (2000): 235.
39   Butler Miller et al. (2008): 399.
40   House of Lords (2006b): para. 6.
41   For a useful review of the critical issues which the notion of 'joined-up government' raises, see Pollitt (2003).
42   Dean (2013): 2.

Chapter 6: Life, death and philosophy

1. Foucault (1991): 143.
2. For introduction to Friedrich Nietzsche, Henri Bergson and Martin Heidegger, particularly with reference to the importance of biology, see Krell (1992); Moore (2002); Ansell Pearson (2002).
3. Foucault (1994b): 387
4. Deleuze (1988): 152.
5. For a discussion of developments within anti-humanism that are important to Michel Foucault and Gilles Deleuze's intellectual formation, see Geroulanos (2010): 287–303.
6. Marx (2000). On Gilles Deleuze and Marxist thought, see Jain (2009).
7. For an introduction to Rosi Braidotti's relationship to Gilles Deleuze, see Braidotti (2012).
8. Braidotti (2007).
9. As observed earlier, when Rosi Braidotti distinguishes between philosophers who offer a negative and an affirmative understanding of life, she situates Michel Foucault among the former, so aligning him with both Martin Heidegger and Giorgio Agamben; see Braidotti (2006): 36–42. For a recent discussion of the relationship between Heidegger and Foucault's philosophical thought, see Rayner (2007).
10. On George Simmel and Sigmund Freud, see also Carel (2006); Pyyhtinen (2012). For a useful introduction to Martin Heidegger's understanding of death and its relationship to his philosophical project, see Schumacher (2011). The discussion of contemporary thought about death, understood as an objective phenomenon coterminous with life, is especially useful; see pp. 131–6. The argument advanced here is, however, at odds with Schumacher's insofar as it proposes that, although not determining, biological considerations about the objective characteristics of ageing and death are as important in shaping Heidegger's thought as any philosophical considerations. In other words, the aim here is not to collapse the distinction between ontic and ontological orders, and to thus fall back on a naturalist interpretation, but to draw attention to the possibility of a more open reading of the relationship between the two orders, so as to at least acknowledge the provocations responsible for these distinctions. While Raoni Padui (2013) discusses the issue in very useful and resonant terms, see also Schumacher (2011): 62–4.

11   Heidegger (1962): 294.
12   See Heidegger (1962): 494. Eugen Korschelt's contributions to biology have gone largely unnoticed, but Jane Maienschein (2009) offers a useful reminder not just of these contributions, but also of the proximity between Korschelt's interest in the biology of ageing and the emergence of regenerative medicine.
13   As noted earlier, Thomas Kessel discusses at great length Martin Heidegger's formative interest in the debates between Wilhelm Roux, Hans Driesch and other contemporaries over the relationship between the organism and its component parts. It is unfortunate that, despite Eugen Korschelt's evident participation in the same debates, no consideration is paid to the potential importance of Korschelt's understanding of the relationship between integration and the organism's mortality for the formation of Heidegger's own understanding of mortality. See particularly Kessel (2011): 207–9.
14   Krell (1992): 33–63. See also Moore (2002) and Bernasconi (2012).
15   Heidegger (1962): 291.
16   See Calarco (2008). While it is widely agreed that Martin Heidegger's understanding of humans and other animals is prey to unwarranted anthropocentrism, William McNeill (1999) offers are a more open reading of Heidegger's distinction between humans and other animals by drawing attention to the manner in which the distinctions at issue are sometimes relative to broader considerations and thus are not to be regarded as categorical.
17   Korschelt (1922): 8 and 413.
18   On Michel Foucault and phenonenology, see May (2003).
19   For an extensive discussion of Michel Foucault's understanding of finitude and its relationship to Martin Heidegger's, see Han (2002): 38–69.
20   For a comprehensive study of Michel Foucault's thought about the historical transformation of reason, from *History of Madness* (1961) onward, see Gutting (1989).
21   The exact interpretation of the 'death of Man' has been the subject of much discussion; see Canguilhem (2003); Han-Pile (2010).
22   Foucault (1994a): 144.
23   Foucault (1994a): 144.
24   Foucault (1991): 138–9.
25   Foucault (1994b): 326.
26   See, for example, Beatrice Han's *Michel Foucault's Critical Project*

(2002), which offers an extensive discussion of the difficulties involved in Foucault's historicizing project. For an early formulation of the same criticism, see instead Derrida (1978).

27 Michel Foucault acknowledged that Xavier Bichat's understanding of the relationship between life and death can be understood as ground in some form of vitalism. Foucault wrote: 'there is much that might be said about Bichat's "vitalism". It is true that in trying to circumscribe the special character of the living phenomenon Bichat linked to its specificity the risk of disease: a simply physical body cannot deviate from its natural type. But this does not alter the fact that the analysis of disease can be carried out only from the point of view of death – of death which life, by definition, resists'; Foucault (1994a): 144. The closing line referred tacitly to François Magendie's critique of Bichat's understanding of the relationship between life and death, which Foucault goes on to discuss immediately after the passage just quoted. As argued elsewhere, it is difficult to reconcile these statements with Gilles Deleuze's notion that 'when Foucault analyses Bichat's theories', his tone demonstrates sufficiently that he is concerned with 'something other than an epistemological analysis'; Deleuze (1988): 95. See also Palladino (2013). Unlike Deleuze, Foucault seemed less enthralled by Bichat's vitalism and more interested in the manifold implications of Bichat's reconfiguration of death as a temporally extended process that was coterminous with the life of anatomical tissues and organs.

28 See Braidotti (2006): 36–42; also Fraser, Kember and Lury (2006). For an introduction to the attempts to reconfigure the understanding of biopolitical governmentality in a manner whereby death and finitude are less important than Michel Foucault's own understanding will allow, see Esposito (2008) and Campbell (2011).

29 Deleuze (1988): 121. While sometimes critical, the relationship between Gilles Deleuze and Martin Heidegger also is very complicated and a number of philosophers are today seeking to clarify the points of contact between the two; see de Beistegui (2004).

30 For a germane introduction to the distinctive nature of Gilles Deleuze's understanding of desire, see Brent Adkins, *Death and Desire in Hegel, Heidegger and Deleuze* (2007). The distinctiveness of Deleuze's construction is evinced by Jacques Lacan's very different response to the implications of the same Hegelian and Heideggerian heritage; see Adkins (2007): 170–95.

31 Deleuze and Guattari (1984): 158.

32 Mark Hansen regards Keith Ansell Pearson's emphasis on the

importance of August Weismann to Gilles Deleuze's thought as 'much exaggerated'; Hansen (2000): note 24. At the same time, Hansen also argues very helpfully that Deleuze and Félix Guattari overstate the insignificance of the organism and that this is related, at least in part, to their reading and elision of differences between Weismann and Henri Bergson. Hansen writes: 'While Weismann and the modern genetics tradition get things wrong by attributing too much to germ cells (genes), D+G [Deleuze and Guattari] go astray by dissolving all ties between germinal continuity and organic life. The error, in both cases, stems from neglect of the organism … There can … be no continuity of germinal life – whether at the level of the germ-line or the plane of immanence – without the active mediation of the organism. The biological can no more be reduced to molecular processes than it can be dissolved through a broader conception of nonorganic life'; Hansen (2000): para. 52. For a more general discussion of the relationship between Deleuze, Guattari and contemporary biology, see Protevi (2012).
33 Bergson (1998): 28–9.
34 Bergson (1935): 45. On the enduring importance of *Two Sources of Morality and Religion*, see Lefebvre and White (2012).
35 Bergson (1935): 107.
36 Bergson (1935): 108.
37 Bergson (1935): 108–9.
38 Bergson (1935): 109.
39 Bergson (1998): 270–1.
40 For an incisive discussion of the way in which Gilles Deleuze would appear to overlook Henri Bergson's critique of Darwinian understanding of natural selection by systematically ignoring the negative implication of selective mechanisms, see Caygill (1997).
41 Deleuze (1991): 42; see also 38–47. For a helpful discussion of the pivotal distinction between the ideal and the virtual, see Delanda (2012): 225–8; see also Ansell Pearson (2002, 2005). Significantly, Keith Ansell Pearson (2005) argues that contemporary constructions of the virtual have drifted away from Henri Bergson's understanding of the relationship between the virtual, the real, the part and the whole. On contemporary understanding, the part has become the real and the whole the virtual, while, according to Ansell Pearson, Bergson regarded the whole as the real and the part as the virtual. Ansell Pearson maintains that Gilles Deleuze remains faithful to this latter understanding, albeit evacuated of all Bergson's subjectivism.

42 As John Marks (2006) has observed, François Jacob and Jacques Monod's work on the regulation of genetic processes informs Gilles Deleuze's understanding of such processes. Since this work famously conjoined genetic and physiological rationalities, it may help to understand Deleuze's distinctive reading of August Weismann's *Essays Upon Heredity and Kindred Biological Problems*.

43 Deleuze (2004): 263.

44 Deleuze (2004): 319.

45 Deleuze (1990): 101. Antonin Artaud coined the notion of 'the body without organs', in the play *To Have Done with the Judgment of God* (1947); see Sontag (1988).

46 Keith Ansell Pearson has observed how Sigmund Freud claimed that his notion of a death drive was attuned to August Weismann's understanding of the relationship between life and death, but Freud's understanding of the death drive actually brings him into conflict with the Weismannian reading of Weismann because it privileged the organism. According to Ansell Pearson, Deleuze's understanding is closer to the Weismannian insistence that the play of life and death operates at the molecular level; see Ansell Pearson (1999): 104–14. For a more general introduction to *The Logic of Sense* and the critique of psychoanalysis advanced therein, see Williams (2008): 175–201.

47 Bousquet, as quoted in Deleuze (1990): 170.

48 Deleuze (1990): 170.

49 Deleuze (1990): 362.

50 This issue is important to the disagreement between Keith Ansell Pearson and Mark Hansen over Gilles Deleuze and Félix Guattari's relationship to contemporary biological thought. While Ansell Pearson emphasizes the turn to August Weismann as facilitating the negotiation between the organism on the one hand, and the molecular and systemic levels of analysis on the other hand, Hansen regards Deleuze and Guattari as committed primarily to the articulation of a novel form of philosophical monism, at the cost of any credible grounding in contemporary biological thought. Hansen concludes his argument by writing that 'organs, the organism, and the assemblage simply cannot be reduced to the philosophical roles D+G [Deleuze and Guattari] reserve for them'; Hansen (2000): para. 66. Hansen's argument is powerful and consistent with the argument advanced here, but Hansen would also appear to overlook the importance of contemporary disagreement about how best

to understand the relationship between the three terms at issue, rearticulated here as the relationship between molecules, populations and the mortal organism. Furthermore, as the next section will seek to establish, Deleuze and Guattari's argument is perhaps better characterized as ambiguous, rather than inconsistent.

51 Deleuze and Guattari (1988): 158.
52 Deleuze and Guattari (1988): 238–9.
53 Deleuze and Guattari (1988): 241.
54 Pepper and Herron (2008).
55 Griesemer (2005): 63.
56 See Griesemer (2005, 2011). Significantly, as part of a long-standing interest in practices of abstraction within biology, James Griesemer has charted the ways in which August Weismann's original understanding of the relationship between somatic and germinal lines has been reconfigured since its first diagrammatic representation, thus reaffirming the ambiguities of all reference to Weismann; see Griesemer and Wimsatt (1989).
57 Deleuze (2004): 236.
58 Griesemer (2011): 38.
59 For a useful introduction to the relationship between continental and analytical philosophy, see Glendinning (2006). Mullarkey (2006) also attends to the fraught relationship between the two traditions, but in addition explores how Deleuzian thought might offer a space for some reconciliation.
60 See also Canguilhem (1991).
61 Interestingly, in her analysis of Deleuzian thought, Claire Colebrook (2006) draws attention to the Deleuzian equation of death and the quantitative numeral zero; see also Deleuze and Guattari (1984): 330. On the peculiarity of the figure zero, pivotal figure of mathematical reason, see Rotman (1993). See also Manuel Delanda (2012) for a discussion of the relationship between mathematics and Deleuzian metaphysics.
62 Foucault (1994b): 326.
63 In *The Parasite* (1980), Michel Serres attempts to rethink the nature of exchange by arguing that it is logically and practically preceded by unidirectional, asymmetric transactions. He introduces the figure of the 'parasite' to foreground the critical importance of disturbances and obscurities to the constitution of any system of exchange. The parasite is that which eats alongside, uninvited, and does not offer anything in exchange. It eats alongside and is

not interested in wherefrom the food came. It eats, and because it was not invited, it has no obligations to either the guest or its host. At the same time, the parasite is that which enables intra-systemic relations insofar as it continuously creates a default in the mutuality of the relations between host and guest, and, as such, is the source of innovation, movement and change. From this perspective, the parasite is both relation and non-relation, because it does not establish a relation with any element of the system and yet enables intra-systemic relations. To remain both in relation and non-relation, however, the parasite has to remain at the boundary of the system of exchanges, always outside to be inside. If it is included, it is enrolled into the system of exchanges and the thus enlarged system will then tend toward equilibrium and eventual cessation of systemic activity. If it is excluded, it will cease to generate difference within the system, so that the flows within the system of exchange will again tend towards equilibrium and eventual cessation of systemic activity. The importance of the parasite thus is that, because it cannot but remain something unaccountable, it creates the very opening for the logic of exchange and calculation. See also Palladino and Moreira (2006). Death might be regarded as a figure comparable to the parasite, being very different to its Derridean configuration as constitutive absence and still very similar insofar as its generative effects. Such combination of conceptual proximity and distance recalls Daniel Smith's discussion of the relationship between Deleuze's and Derrida's understanding of difference and differentiation; Smith (2012).

Chapter 7: The arts of living and dying

1   See also Cooper (2009) and Palladino (2013).
2   Deleuze, as quoted in Agamben (1999): 228–9; see also Deleuze (2001): 28–9.
3   Deleuze, as quoted in Agamben (1999): 228; see also Deleuze and Guattari (1994): 40.
4   Dickens, as quoted in Agamben (1999): 229; see also Dickens (1989): 444–5.
5   Agamben (1999): 230.
6   Cooper (2009): 158.
7   Mitchell Dean evokes the proximity of his and Melinda Cooper's argument when discussing the disappearance of sovereign power from

contemporary reflections on the conjunction of biomedical enterprise and neo-liberal governance; Dean (2013): 93.
8   Foucault (1991): 143.
9   Foucault (2004): 256–60.
10 Foucault (2004): 248–9.
11 Foucault (2004): 240.
12 Agamben (1998): 162.
13 Agamben (1998): 163–4.
14 Agamben (1998): 186.
15 Agamben (1998): 187. Contrast Giorgio Agamben's conclusion with Michel Foucault's evocation of 'bodies and pleasures' as the the potential site of emancipation; Foucault (1991): 157.
16 Seale (1998): 91–121.
17 For an important articulation of the difference between the two perspectives on life, see Singer (1994): 106–31. Furthermore, Shershow (2012, 2014) discusses very usefully the contemporary debates over the right to die, bringing Derridean deconstruction to bear on the pivotal distinction between the sanctity and dignity of life, arguing that it takes considerable dexterity to keep them from blurring into one another, partly because the concept of dignity is internally riven and in a way that undoes the distinction.
18 Hanafin (2008): 55; see also Deleuze and Guattari (1988): 279–80.
19 Hanafin (2008): 57.
20 As discussed below, Giorgio Agamben attends to Alexandre Kojève's reflections on Japanese ritual suicide. He does not, however, draw any connection between this gesture and his ensuing reflections on life, the body and the end of history; Agamben (2004): 9–12. On suicide and its evolving relationship to juridico-political formations, see Minois (1999).
21 For an early acknowledgement of Rabinow's debt to Foucault, see Rabinow (1989): ix–x.
22 Rabinow (1999a): 174.
23 Rabinow and Rose (2003): 3; see also Rabinow and Rose (2006). Giorgio Agamben regards Paul Rabinow's notion of an 'experimental' mode of inquiry as complicit with the transgressions of law and morality which characterize contemporary biopolitical governmentality; see Agamben (1998): 185–6.
24 Rabinow (2003): 61–8.

25 See Rabinow (2008): 13–14; Taubes (2010): 115–18.
26 Rabinow (2008): 101–28.
27 Taubes (2010): 99.
28 Taubes (2010): 103.
29 Taubes (2010): 110.
30 Rabinow (2008): 112.
31 Foucault, as quoted in Campbell (2011): 134; see also Foucault (2005): 486.
32 Campbell (2011): 134.
33 On Agamben and play, see also Mills (2008).
34 Stelarc, as quoted in Caygill (2000): 162.
35 Deleuze (1990): 226–7.
36 On the ambiguities of monsters and their productivity, see Cooper (2008b).
37 Seale (1998): 144–5.
38 For a discussion of the case and its historical significance, see Warnock and Macdonald (2008). The distinction between formal and substantive relationship is drawn from Giorgio Agamben's discussion of Alexandre Kojève's amendments to his lectures on G. W. F. Hegel's *Phenomenology of Spirit* (1807); Agamben (2004): 9–12. By referring to this formative text, Agamben's aim is to draw out how anti-humanism is bound to remain in the thrall of idealism so long as it fails to inquire into the genesis of the opposition between the human and the animal, and he includes Michel Foucault among those at fault. He is interested in Kojève's amendments because they aimed to revise his earlier account of the 'end of history' in the light of reflections on the formal culture of modern Japan, and focused particularly on *seppuku*, ritual disembowelment, which, according to Kojève, had become no more than a 'perfectly gratuitous suicide, which has nothing to do with the risk of life in a fight waged for the sake of historical values'; see Kojève (1980): 159–62.
39 Dyer (2002).
40 See Singer (1994): 57–80; Keown (2002): 217–81.
41 Select Committee on Medical Ethics, as cited in House of Lords (2001): para. 29.
42 House of Lords (2001): para. 29.
43 See Gilchrist (2000). For a discussion of the global reach of the phenomenon, see Cherniack (2002).

44 Strikingly, while the Law Lords' articulation of the potential impact of any decriminalization of assisted suicide upon the elderly focuses on coercion, it overlooks how contemporary constructions of ageing and the aged elide the boundaries drawn between voluntary and coerced requests. As Harry Moody observes in his reflecting on the notion of successful ageing, which contemporary biogerontology seeks to secure: 'if the aim of life is to be successful, and if success is to be measured by life satisfaction, then when homeostasis and quality of life decline below a certain point, it's time for assisted suicide'; Moody (2001): 179. See also Shershow (2014).
45 Dyer (2001).
46 Barclay (2002).
47 Charis Thompson discusses very usefully the ontological effects of such choreography; see Thompson (2005). Thompson's debt to Rosi Braidotti's *Nomadic Subjects* (1994) is considerable.
48 Deleuze and Guattari, as quoted in Protevi (2008): 71; see also Deleuze and Guattari (1988): 35.
49 Deleuze and Guattari (1988): 36.
50 Scott Shershow also examines John Protevi's discussion of Terry Schiavo's case, albeit with reference to Protevi's discussion in *Political Affect* (2009), coming to a similar conclusion as that advanced here; Shershow (2014): 158–61.
51 Such entanglement of critical practices in the present lies at the heart of Mitchell Dean's analysis of power in *The Signature of Power* (2013).
52 Bousquet, as quoted in Deleuze (1990): 170.
53 Foucault (2005): 477.
54 Marcus Aurelius, as quoted in Foucault (2005): 478.
55 Foucault (2005): 487; see also Foucault (2000).
56 See Pierre Hadot's criticism of Michel Foucault's reading of Stoicism as prey to 'dandyism', in Hadot (1995).
57 Agamben (2004): 83.
58 Agamben (2004): 86–7.
59 Žižek (1999): 262; see also Palladino (2010).
60 Bousquet, as quoted in Deleuze (1990): 170; see also Smith (2004).
61 Nietzsche (1911): 250.
62 How best to understand Friedrich Nietzsche's notion of the eternal return, how to affirm life and yet evacuate it of all possibility of

return and any subterranean notion of life eternal, is difficult. As Didier Franck has observed recently, Nietzsche's characterization of the eternal return changed substantially, from an initial injunction to 'live in such a way that you would desire to live anew, that is the task – you will live anew in any case' to a later understanding that the task is 'to live [so] that there is no longer any meaning in living: that now becomes the "meaning" of life'; Franck (2012): 337. The thrust here is to advance the latter understanding.

Conclusion

1 Gormley n.d.
2 Taubes (2009): 3–4.
3 Taubes (2009): 191.
4 Jonas (2001): 12.
5 See Agamben (2004): 83.
6 Deleuze (1988): 132; see also Canguilhem (2003).
7 Agamben (2004): 92; see also Oliver (2007).
8 Braidotti (2007).
9 See also Pickstone (2011). As attests John Law's *Organizing Modernity* (1993), the attention John Pickstone devotes to heterogeneous ordering is neither novel, nor unique, but Pickstone's displacement of sequential ordering draws much needed attention to the often unspoken historiographical urge to reduce diversity into an orderly and uniform temporal procession.
10 The difficulties involved in holding to Michel Foucault's insistence on the systemic nature of discursive formations and the simultaneous refusal of periodization are evident in *The Archaeology of Knowledge* (1969); see Foucault (1972): 16–17 and 148.
11 Cf. Pickstone (2001): 41–5.
12 Rose (2007): 223.
13 Katz and Marshall (2003): 5.
14 Keating and Cambrosio (2003): 20.
15 Yeats (1950): 211.
16 Rabinow (2008): 49.
17 Rabinow (2008): 49–50.
18 The notion of parasitism deployed here does not point to any

illegitimacy, but to the complex relationship that Michel Serres conjures in *Le Parasite* (1980); see also Palladino and Moreira (2006).

19 Foucault (1977): 146.

20 On the importance of aesthetics to Gilles Deleuze's argument, see Olkowski (2012). While Paul Rabinow's recent work, which, as observed in previous chapters, attends closely to the resonance between bio-art and contemporary biomedical and life science, calls for a similar analysis, his aesthetics still await sustained analysis.

21 Boltanski and Thévenot (2006): 160–1.

22 Admittedly, John Law's *Organizing Modernity* (1993) offers much the same insight and in a manner closer to the interest in socio-technical transformations, but Luc Boltanski and Laurent Thévenot offer the added advantage of being interested in formulating a sociology of critical capacities, as well as deploying a resonant mode of representation. This is not to say that Law is not interested in the articulation of alternative critical capacities, but that there is no comparable attention to the role of multiplicity in the constitution of the subject, riven, rather than hybrid; see Law (2004). This last contrast is critical to the importance attached here to Boltanski and Thévenot's *On Justification*; see also Guggenheim and Potthast (2012).

23 The attributes and value characterizing the 'polity of fame' are outlined on pp. 98–107. If Luc Boltanski and Laurent Thévenot's construction of the metaphysics of power might then seem to rest on the model of sovereignty, pastoral power also finds a place in their analysis, under the guise of Jacques-Bénigne Bossuet's 'domestic polity'; see pp. 90–8.

24 See Boltanski and Thévenot (2006): 80–2, 343.

25 For an earlier discussion of the structural and tropological features of such an argument, see Palladino (2007). The challenge is to resist all denunciatory modes.

26 See Serres (1980): 128; also Ricoeur (2005): 201–16 and 220–5. I suppose that, in the last analysis and in at least in some ways, I find myself agreeing with Alain Badiou (1997) and his argument that Deleuzian thought reaffirms the very metaphysics that it seeks to overcome.

WORKS CITED

Achenbaum, Andrew W. *Crossing Frontiers: Gerontology Emerges as a Science*. Cambridge: Cambridge University Press, 1995.
Adkins, Brent. *Death and Desire in Hegel, Heidegger and Deleuze*. Edinburgh: Edinburgh University Press, 2007.
Agamben, Giorgio. *Homo Sacer: Sovereign Power and Bare Life [1995]*. Stanford: Stanford University Press, 1998.
Agamben, Giorgio. 'Absolute immanence [1996]'. In *Potentialities: Collected Essays in Philosophy*, 220–39. Stanford: Stanford University Press, 1999.
Agamben, Giorgio. *The Open: Man and Animal [2002]*. Stanford: Stanford University Press, 2004.
Agamben, Giorgio. *The Time That Remains: A Commentary on the Letter to the Romans [2000]*. Stanford: Stanford University Press, 2005.
Agamben, Giorgio. *The Signature of All Things: On Method*. New York: Zone Books, 2009.
Agamben, Giorgio. *The Kingdom and the Glory: For a Theological Genealogy of Economy and Government*. Stanford: Stanford University Press, 2011.
Agamben, Giorgio. *Nudities [2009]*. Stanford: Stanford University Press, 2011.
Agar, Nicholas. *Humanity's End: Why We Should Reject Radical Enhancement*. Cambridge, MA: MIT Press, 2010.
Alliance for Aging Research. 'The Science of Aging Gracefully: Scientists and the Public Talk about Aging Research'. 2005. Available online: http://www. publicagenda. org/files/science_of_aging_gracefully. pdf (accessed 20 March 2013).
Amsterdamska, Olga. 'Demarcating Epidemiology'. *Science, Technology & Human Values* 30 (2005): 17–51.
Anidjar, Gil. 'The Meaning of Life'. *Critical Inquiry* 37 (2011): 697–723.
Ansell Pearson, Keith. *Germinal Life: The Difference and Repetition of Deleuze*. London: Routledge, 1999.

Ansell Pearson, Keith. *Philosophy and the Adventure of the Virtual: Bergson and the Time of Life*. London: Routledge, 2002.
Ansell Pearson, Keith. 'The Reality of the Virtual: Bergson and Deleuze'. *MLN* 120 (5) (2005): 1112–27.
Ariés, Philippe. *Hours of Our Death [1977]*. Harmondsworth: Penguin, 1987.
Armstrong, David. *Political Anatomy of the Body: Medical Knowledge in Britain in the Twentieth Century*. Cambridge: Cambridge University Press, 1983.
Asch, Adrienne. 'Big Tent Bioethics: Toward an Inclusive and Reasonable Bioethics'. *The Hastings Center Report* 35 (6) (2005): 11–12.
Assmann, Aleida, Jan Assmann and Wolf-Daniel Hartwich. 'Introduction to the German edition [1996]'. In *From Cult to Culture: Fragments Toward a Critique of Historical Reason*, edited by Charlotte Elisheva Fonrobert and Amir Engel, xviii–l. Stanford: Stanford University Press, 2010.
Atkinson, Paul. *Medical Talk and Medical Work*. London: Sage, 1995.
Badiou, Alain. *Deleuze: The Clamor of Being [1997]*. Minneapolis: University of Minnesota Press, 2000.
Ballenger, Jesse. *Self, Senility, and Alzheimer's Disease in Modern America*. Baltimore: Johns Hopkins University Press, 2006.
Barclay, Sarah. 'It's Not Life. I'm Already Dead'. 12 May 2002. Available online: http://www. theguardian. com/theobserver/2002/may/12/featuresreview.review (accessed 20 February 2014).
Bauman, Zygmunt. *Mortality, Immortality & Other Life Strategies*. Stanford: Stanford University Press, 1992.
Bauman, Zygmunt. *Life in Fragments: Essays in Postmodern Morality*. Oxford: Blackwell, 1995.
Bayatrizi, Zoreh. *Life Sentences: The Modern Ordering of Mortality*. Toronto: University of Toronto Press, 2008.
Beistegui, Miguel de. *Truth and Genesis: Philosophy as Differential Ontology*. Bloomington: Indiana University Press, 2004.
Bergson, Henri. *The Two Sources of Morality and Religion [1932]*. London: MacMillan, 1935.
Bergson, Henri. *Creative Evolution [1911]*. New York: Dover, 1998.
Bernard, Claude. *Introduction to the Study of Experimental Medicine [1865]*. New York: Dover, 1957.
Bernasconi, Robert. 'Heidegger, Rickert, Nietzsche, and the Question of Biologism'. In *Heidegger & Nietzsche*, edited by Babette E. Babich, Alfred Denker and Holger Zaborowski, 159–80. Amsterdam: Rodopi, 2012.
Bichat, Xavier. *Recherches Physiologiques sur la Vie et la Mort [1800]*. Paris: Flammarion, 1994.

Blokker, Paul. 'Pragmatic Sociology: Theoretical Evolvement and Empirical Application'. *European Journal of Social Theory* 14 (3) (2011): 251–61.

Boltanski, Luc and Laurent Thévenot. *On Justification: Economies of Worth [1991]*. Princeton: Princeton University Press, 2006.

Bookchin, Debbie and Jim Schumacher. *The Virus and the Vaccine: Contaminated Vaccine, Deadly Cancers, and Government Neglect*. New York: St. Martin's Press, 2005.

Borch-Jacobsen, Mikkel. *Lacan: The Absolute Master*. Stanford: Stanford University Press, 1991.

Boyne, Roy. 'Angels in the Archive: Lines into the Future in the Work of Jacques Derrida and Michel Serres'. *Cultural Values* 2 (2) (1998): 206–22.

Braidotti, Rosi. *Nomadic Subjects: Embodiment and Sexual Difference in Contemporary Feminist Theory*. New York: Columbia University Press, 1994.

Braidotti, Rosi. *Metamorphoses: Towards a Materialist Theory of Becoming*. Cambridge: Polity Press, 2002.

Braidotti, Rosi. *Transpositions: On Nomadic Ethics*. Cambridge: Polity, 2006.

Braidotti, Rosi. 'Biopower and Necro-politics: Reflections on an Ethics of Sustainability'. *Springerin*. 2007.

Braidotti, Rosi. 'Nomadic Ethics'. In *The Cambridge Companion to Deleuze*, edited by Daniel W. Smith and Henry Somers-Hall, 170–97. Cambridge: Cambridge University Press, 2012.

British Broadcasting Corporation. 'BBC Radio 4 Unveils 60 Years of Reith Lectures Archive'. 26 June 2011. Available online: http://www.bbc.co.uk/news/entertainment-arts-13891740 (accessed 4 September 2013).

Brown, Steve. 'Michel Serres: Science, Translation and the Logic of the Parasite'. *Theory, Culture and Society* 19 (3) (2002): 1–27.

Bryant, Levi, Nick Srnicek, and Graham Harman. 'Towards a Speculative Philosophy'. In *The Speculative Turn: Continental Materialism and Realism*, edited by Levi Bryant, Nick Srnicek and Graham Harman, 1–18. Victoria: re.press, 2011.

Butler, Judith. *Bodies that Matter: On the Discursive Limits of Sex*. London: Routledge, 1993.

Butler, Robert N. *Why Survive? Growing Old in America*. New York: Harper & Row, 1975.

Butler, Robert N. 'Research Programs of the National Institute of Aging'. *Public Health Reports* 92 (1) (1977): 3–8.

Butler, Robert N. *The Longevity Revolution: The Benefits and Challenges of Living a Long Life*. New York: Public Affairs, 2008.

Butler, Robert N., Richard A. Miller, Daniel Perry, Bruce A. Carnes, Franklin T. Williams, Christine Cassell, Jacob Brody, Jacob, Marie A. Bernard, Linda Partridge, Thomas Kirkwood, George Martin and Jay S. Olshansky. 'New Model of Health Promotion and Disease Prevention for the 21st Century'. *British Medical Journal* 337 (2008): 399.

Cabinet Office. 'Horizon Scanning Programme: A New Approach for Policy Making'. 12 July 2013. Available online: https://www. gov. uk/government/news/horizon-scanning-programme-a-new-approach-for-policy-making (accessed 23 September 2013).

Calarco, Matthew. *Zoographies: The Question of the Animal from Heidegger to Derrida*. New York: Columbia University Press, 2008.

Cambrosio, Alberto, Peter Keating, Thomas Schlich, and George Weisz. 'Biomedical Conventions and Regulatory Objectivity: A Few Introductory Remarks'. *Social Studies of Science* 39 (5) (2009): 651–64.

Campbell, Timothy C. *Improper Life: Technology and Biopolitics from Heidegger to Agamben*. Minneapolis: University of Minnesota Press, 2011.

Canguilhem, Georges. *The Normal and the Pathological [1966]*. New York: Zone Books, 1991.

Canguilhem, Georges. 'The Death of Man, or the Exhaustion of the Cogito?' In *The Cambridge Companion to Foucault*, edited by Gary Gutting, 74–94. Cambridge: Cambridge University Press, 2003.

Carel, Havi. *Life and Death in Freud and Heidegger*. Amsterdam: Rodopi, 2006.

Carroll, Lewis. '*Alice's Adventures in Wonderland* [1865]. Project Gutenberg.' 4 November 2012. Available online: http://www.gutenberg.org/files/11/11-h/11-h. htm (accessed 24 March 2014).

Carroll, Lewis. '*Through the Looking Glass* [1871]. Project Gutenberg.' 8 January 2013. Available online: http://www.gutenberg.org/files/12/12-h/12-h.htm (accessed 24 March 2014).

Caygill, Howard. 'The Topology of Selection'. In *Deleuze and Philosophy: The Difference Engineer*, edited by Keith Ansell Pearson, 149–62. London: Routledge, 1997.

Caygill, Howard. 'Liturgies of Fear: Biotechnology and Culture'. In *The Risk Society and Beyond: Critical Issues for Social Theory*, edited by Barbara Adam, Ulrich Beck and Joost Van Loon, 155–64. London: Sage, 2000.

Chadarevian, Soraya de and Harmke Kamminga. 'Introduction'. In *Molecularizing Biology and Medicine: New Practices and Alliances, 1910s–1970s*, edited by Soraya de Chadarevian and Harmke Kamminga, 1–16. Amsterdam: Harwood Academic Press, 1998.

Charise, Andrea. '"Let the Reader Think of the Burden": Old Age and the Crisis of Capacity'. *Occasion* (2012).
Cherniack, Evan P. 'Increasing Use of DNR Orders in the Elderly Worldwide: Whose Choice is it?' *Journal of Medical Ethics* 28 (2002): 303–7.
Churchill, Frederick. 'From Heredity Theory to Vererbung: The Transmission Problem, 1850–1915'. *Isis* 78 (1987): 336–64.
Clarke, Adele E., Laura Mamo, Jennifer Ruth Fosket, Jennifer R. Fishman and Janet K. Shim. 'Biomedicalization: Technoscientific Transformations of Health, Illness, and U.S. Biomedicine'. *American Sociological Review* 68 (2003): 161–94.
Cohen, Lawrence. *No Aging in India: Alzheimer's, the Bad Family, and other Modern Things*. Berkeley: University of California Press, 2000.
Colebrook, Claire. *Deleuze: A Guide for the Perplexed*. London: Bloomsbury, 2006.
Colgrove, James. 'The McKeown Thesis: A Historical Controversy and its Enduring Influence'. *American Journal of Public Health* 92 (2002): 725–9.
Comfort, Alex. *Sex in Society*. London: Duckworth, 1950.
Comfort, Alex. *The Biology of Senescence*. London: Routledge, Kegan & Paul, 1956.
Comfort, Alex. 'Ageing in Animals'. In *Symposia of the Institute of Biology, No. 6: The Biology of Ageing*, edited by William Brunsdon Yapp and Geoffrey Howard Bourne, 23–6. London: Haefner, 1957.
Comfort, Alex. *The Joy of Sex: A Gourmet Guide to Lovemaking*. London: Quartet, 1972.
Comfort, Alex. *I and That: Notes on the Biology of Religion*. New York: Crown Publishers, 1979.
Comfort, Alex. *Reality & Empathy: Physics, Mind and Science in the 21st Century*. Albany: State University of New York Press, 1984.
Cooper, Melinda. 'Resuscitations: Stem Cells and the Crisis of Old Age'. *Body & Society* 12 (2006): 1–23.
Cooper, Melinda. *Life as Surplus: Biotechnology and Capitalism in the Neoliberal Era*. Seattle: University of Washington Press, 2008a.
Cooper, Melinda. 'Monstrous Progeny: The Teratological Tradition in Science and Literature'. In *Frankenstein's Science, Experimentation and Discovery in Romantic Culture, 1780-1830*, edited by Christa Knellwolf and Jane Goodall, 87–98. Aldershot: Ashgate, 2008b.
Cooper, Melinda. 'The Silent Scream: Agamben, Deleuze and the Politics of the Unborn'. In *Deleuze and Law: Forensic Futures*, edited by Rosi Braidotti, Claire Colebrook and Patrick Hanafin, 142–62. London: Palgrave Macmillan, 2009.

Cowdry, Edmund V., ed. *Problems of Ageing*. Baltimore: Williams & Wilkins, 1939.

Cowdry, Edmund V., ed. 'Tissue Fluids'. In *Problems of Ageing*, edited by Edmund Cowdry, 642–94. Baltimore: Williams & Wilkins, 1939.

Creager, Angela N. H. *The Life of a Virus: Tobacco Mosaic Virus as an Experimental Model 1930–1965*. Chicago: University of Chicago Press, 2002.

Cross, Stephen T. and Randall W. Albury. 'Walter B. Cannon, L. J. Henderson, and the Organic Analogy'. *Osiris* 3 (1987): 165–92.

Currier, Dianne. 'Feminist Technological Futures: Deleuze and Body/Technology Assemblages'. *Feminist Theory* 4 (3) (2003): 321–38.

Darwin, Charles. *On the Origin of Species [1859]*. Cambridge: Harvard University Press, 1964.

Davis, Kathleen. *Periodization and Sovereignty: How Ideas of Feudalism and Secularization Govern the Politics of Time*. Philadelphia: University of Pennsylvania Press, 2008.

Dawkins, Richard. *The Selfish Gene*. Oxford: Oxford University Press, 1976.

Dean, Mitchell. *Critical and Effective Histories: Foucault's Methods and Historical Sociology*. London: Routledge, 1994.

Dean, Mitchell. *Governmentality: Power and Rule in Modern Society*. Los Angeles: Sage, 1999.

Dean, Mitchell. *The Signature of Power: Sovereignty, Governmentality and Biopolitics*. Los Angeles: Sage, 2013.

Delanda, Manuel. 'Deleuze, Mathematics, and Realist Ontology'. In *The Cambridge Companion to Deleuze*, edited by Daniel W. Smith and Henry Somers-Hall, 220–38. Cambridge: Cambridge University Press, 2012.

Deleuze, Gilles. *Foucault [1986]*. London: Athlone Press, 1988.

Deleuze, Gilles. *The Logic of Sense [1969]*. London: Bloomsbury, 1990.

Deleuze, Gilles. *Bergsonism [1966]*. New York: Zone Books, 1991.

Deleuze, Gilles. 'Postscript on the Societies of Control [1990]'. *October* 59 (1992): 3–7.

Deleuze, Gilles. 'Immanence: A Life … [1995]'. In *Pure Immanence: Essays on A Life*, 25–33. New York: Zone Books, 2001.

Deleuze, Gilles. *Difference and Repetition [1968]*. London: Continuum, 2004.

Deleuze, Gilles and Fèlix Guattari. *Anti-Oedipus: Capitalism and Schizophrenia [1972]*. London: Athlone, 1984.

Deleuze, Gilles and Fèlix Guattari. *A Thousand Plateaus: Capitalism & Schizophrenia [1980]*. London: Continuum, 1988.

Deleuze, Gilles and Fèlix Guattari. *What is Philosophy? [1991]*. London: Verso, 1994.

Department for Work and Pensions. Available online: https://www. gov. uk/government/organisations/department-for-work-pensions (accessed 20 September 2013).

Derrida, Jacques. 'Cogito and the History of Madness [1966]'. In *Writing and Difference*, 31–63. London: Routledge, 1978.

Dickens, Charles. *Our Mutual Friend [1865]*. Oxford: Oxford University Press, 1989.

Dolphijn, Rick and Iris van der Tuin. 'Interview with Rosi Braidotti'. In *New Materialism: Interviews & Cartographies*, edited by Rick Dolphijn and Iris van der Tuin, 19–37. Ann Arbor: University of Michigan Press, 2012.

Dreyfus, Hubert and Paul Rabinow. *Foucault, Beyond Structuralism and Hermeneutics*. Chicago: University of Chicago Press, 1983.

Dromii, Shai and Eva llouzii. 'Recovering Morality: Pragmatic Sociology and Literary Studies'. *New Literary History* 41 (2) (2010): 351–69.

Dyer, Clare. 'Law Lords Reject Right to Die Plea'. 30 November 2001. Available online: http://www. theguardian. com/uk/2001/nov/30/claredyer (accessed 20 February 2014).

Dyer, Clare. '"Free at Last": Diane Pretty Dies'. 13 May 2002. Available online: http://www theguardian. com/society/2002/may/13/health. healthandwellbeing (accessed 20 February 2014).

Edwardson, James A. and Thomas B. L. Kirkwood. 'The Institute of Ageing and Health, University of Newcastle'. *Experimental Gerontology* 37 (2002): 749–56.

Emden, Christian J. *Friedrich Nietzsche and the Politics of History*. Cambridge: Cambridge University Press, 2008.

Esposito, Roberto. *Bios: Biopolitics and Philosophy*. Minneapolis: University of Minnesota Press, 2008.

Esposito, Roberto. *Immunitas: The Protection and Negation of Life*. Cambridge: Polity, 2011.

Estes, Carroll L. and Elizabeth Binney. 'The Biomedicalization of Aging: Dangers and Dilemmas'. *Gerontologist* 29 (1989): 587–96.

European Union. 'European Research Area on Ageing 2.' 2010. Available online: ftp://ftp.cordis.europa.eu/pub/fp7/coordination/docs/eraage2_en.pdf (accessed 29 March 2013).

Flynn, Thomas R. *Sartre, Foucault and Historical Reason: A Poststructuralist Mapping of History*. Chicago: University of Chicago Press, 2005.

Foucault, Michel. *The Archaeology of Knowledge [1969]*. New York: Pantheon, 1972.

Foucault, Michel. 'Nietzsche, Genealogy, History [1971]'. In *Language, Counter-Memory, Practice: Selected Essays and Interviews by Michel*

Foucault, edited by Donald Bouchard, 139–64. Ithaca: Cornell University Press, 1977.
Foucault, Michel. 'Truth and Power [1977]'. In *Power/Knowledge: Selected Interviews and Other Writings, 1972–1977*, edited by Colin Gordon, 109–33. New York: Pantheon, 1980.
Foucault, Michel. *Care of the Self [1984]*. Harmondsworth: Penguin, 1990.
Foucault, Michel. *Discipline and Punish: The Birth of the Prison [1975]*. Harmondsworth: Penguin, 1991.
Foucault, Michel. *The History of Sexuality: An Introduction [1976]*. Harmondsworth: Penguin, 1991.
Foucault, Michel. 'Questions of Method [1980]'. In *The Foucault Effect: Studies in Governmentality*, edited by Graham Burchell, Colin Gordon and Peter Miller, 73–86. Chicago: University of Chicago Press, 1991.
Foucault, Michel. *The Birth of the Clinic: An Archaeology of Medical Perception [1964]*. New York: Vintage Press, 1994a.
Foucault, Michel. *The Order of Things: An Archaeology of the Human Sciences [1966]*. New York: Vintage Press, 1994b.
Foucault, Michel. 'What Our Present Is [1981]'. In *Foucault Live: Collected Interviews, 1961–1984*, edited by Sylvère Lotringer, 407–15. New York: Semiotext(e), 1996.
Foucault, Michel. 'What is Enlightenment? [1984]'. In *Michel Foucault: Essential Works of Foucault, 1954–1984*, edited by Paul Rabinow, 303–19, Vol. 1. Harmondsworth: Penguin, 2000.
Foucault, Michel. *Society Must Be Defended: Lectures at the Collège de France, 1975–1976*. Harmondsworth: Penguin, 2004.
Foucault, Michel. *The Hermeneutics of the Subject: Lectures at the Collège de France, 1981–1982*. New York: Picador, 2005.
Foucault, Michel. *History of Madness [1961]*. Abingdon: Routledge, 2006.
Foucault, Michel. *Security, Territory, Population: Lectures at the Collège de France, 1977–1978*. London: Palgrave Macmillan, 2009.
Foucault, Michel. *The Courage of the Truth: Lectures at the Collège de France, 1983–1984*. London: Palgrave Macmillan, 2011.
Fox, Daniel M. 'The Significance of the Milbank Memorial Fund for Policy: An Assessment at its Centennial'. *Milbank Quarterly* 84 (2006): 5–36.
Fox-Keller, Evelyn. *Making Sense of Life: Explaining Biological Development with Models, Metaphors and Machines*. Cambridge, MA: Harvard University Press, 2002.
Fox-Keller, Evelyn. *The Century of the Gene*. Cambridge, MA: Harvard University Press, 2000.

Franck, Didier. *Nietzsche and the Shadow of God*. Evanston: Northwestern University Press, 2012.
Fraser, Mariam, Sarah Kember and Celia Lury. 'Inventive Life'. In *Inventive Life: Approaches to the New Vitalism*, edited by Mariam Fraser, Sarah Kember and Celia Lury, 1–14. London: Sage, 2006.
Freud, Sigmund. *Beyond the Pleasure Principle [1922]*. New York: Norton, 1990.
Friedman, David M. *The Immortalists: Charles Lindbergh, Dr. Alexis Carrel, and Their Daring Quest to Live Forever*. New York: Harper Collins, 2008.
Fries, James. 'The Compression of Morbidity [1983]'. *Milbank Quarterly* 83 (2005): 801–23.
Fujimura, Joan H. *Crafting Science: A Sociohistory of the Quest for the Genetics of Cancer*. Cambridge: Harvard University Press, 1996.
Gaudillière, Jean-Paul. 'Le cancer entre infection et hérédité: Gènes, virus et souris au National Cancer Institute (1937–1977)'. *Revue d'Historie des Sciences* 47 (1994): 57–89.
Gaudillière, Jean-Paul. 'The Molecularization of Cancer Etiology in the Postwar United States: Instruments, Politics and Management'. In *Molecularizing Biology and Medicine: New Practices and Alliances, 1910s–1970s*, edited by Soraya de Chadarevian and Harmke Kamminga, 139–70. Amsterdam: Harwood Academic, 1998.
Gaudillière, Jean-Paul. *Inventer la Biomédecine: La France, l'Amérique et la Production des Savoirs du Vivant, 1945–1965*. Paris: La Découverte, 2002.
Geroulanos, Stefanos. 'Beyond the *Normal and the Pathological*: Recent Literature on Georges Canguilhem'. *Gesnerus* 66 (2) (2009): 288–306.
Geroulanos, Stefanos. *An Atheism that is not Humanist Emerges in French Thought*. Stanford: Stanford University Press, 2010.
Gigerenzer, Gerd, Zeno Swijtink, Theodore Porter, Lorraine Daston, John Beatty and Lorenz Kruger. *The Empire of Chance: How Probability Changed Science and Everyday Life*. New York: Cambridge University Press, 1989.
Gilchrist, Caroline. 'Too Old to Care'. 17 May 2000. Available online: http://www.theguardian.com/society/2000/may/17/guardiansocietysupplement6 (accessed 11 June 2014).
Glendinning, Simon. *The Idea of Continental Philosophy: A Philosophical Chronicle*. New York: Columbia University Press, 2006.
Goff, Robert. 'The Joy of Aging: Alex Comfort and the Popularization of Gerontology'. In *Narratives of Life: Mediating Age*, edited by Heike Hartung and Roberta Maierhofer, 71–87. Munster: LIT Verlag, 2009.

Goldman, Marlene. 'Aging, Old Age, Memory, Aesthetics: Introduction to Special Issue'. *Occasion*. 2012.

Gormley, Antony. 'Another Place'. 1997. Available online: http://www.antonygormley. com/sculpture/item-view/id/230 (accessed 2 August 2013).

Grey, Aubrey de. '1000-year Lifespans are Closer than they Seem: A Reply to Kirkwood'. 2005a. Available online: http://sageke.sciencemag. org/community/forum/short/sageke_el_324. dtl (accessed 20 March 2013).

Grey, Aubrey de. 'A Roadmap to End Aging'. 2005b. Available online: http://www. ted. com/talks/aubrey_de_grey_says_we_can_avoid_aging.html (accessed 20 March 2013).

Griesemer, James R. 'The Informational Gene and the Substantial Body: On the Generalization of Evolutionary Theory by Abstraction'. In *Idealization XII: Correcting the Model. Idealization and Abstraction in the Sciences*, edited by Jones Martin and Nancy Cartwright, 59-115. Amsterdam: Rodopi, 2005.

Griesemer, James R. 'Heuristic Reductionism and the Relative Significance of Epigenetic Inheritance in Evolution'. In *Linking Genotype and Phenotype in Development and Evolution*, edited by Benedikt Hallgrimsson and Brian K. Hall, 14–40. Berkeley: University of California Press, 2011.

Griesemer, James R. and William Wimsatt. 'Picturing Wesimannism: A Case Study of Conceptual Evolution'. In *What the Philosophy of Biology Is: Essays Dedicated to David Hull*, edited by Michael Ruse, 75–137. Boston: Kluewer Academic Press, 1989.

Griesemer, James R. and Elihu Gerson. 'Of Mice and Men and Low Unit Cost'. *Studies in History & Philosophy of Biological and Biomedical Sciences* 37 (2006): 363–72.

Grosz, Elizabeth. *The Nick of Time: Politics, Evolution and the Untimely*. Durham: Duke University Press, 2004.

Guggenheim, Michael and Jörg Potthast. 'Symmetrical Twins: On the Relationship between Actor-network Theory and the Sociology of Critical Capacities'. *European Journal of Social Theory* 15 (2012): 157–78.

Gutting, Gary. *Michel Foucault's Archaeology of Scientific Reason: Science and the History of Reason*. Cambridge: Cambridge University Press, 1989.

Haber, Carole and Brian Gratton. *Old Age and the Search for Security: An American Social History*. Bloomington: Indiana University Press, 1994.

Hacking, Ian. *The Taming of Chance*. Cambridge: Cambridge University Press, 1990.

Hadot, Pierre. *Philosophy as a Way of Life*. Oxford: Blackwell Publishing, 1995.
Haldane, J. B. S. *The Causes of Evolution*. London: Longman, 1932.
Haldane, J. B. S. *New Paths in Genetics*. London: Allen & Unwin, 1941.
Hall, Stephen. *Merchants of Immortality: Chasing the Dream of Human Life Extension*. Boston: Houghton Mifflin, 2003.
Han, Béatrice. *Michel Foucault's Critical Project: Between the Transcendental and the Historical*. Stanford: Stanford University Press, 2002.
Hanafin, Patrick. 'Rights of Passage: Law and the Bio-politics of Dying'. In *Deleuze and Law: Forensic Futures*, edited by Rosi Braidotti, Claire Colebrook and Patrick Hanafin, 47–58. London: Palgrave Macmillan, 2008.
Han-Pile, Béatrice. 'The "Death of Man": Foucault and Anti-humanism'. In *Foucault and Philosophy*, edited by Christopher Falzon Timothy O'Leary, 118–43. Malden: Wiley-Blackwell, 2010.
Hansen, Mark. 'Becoming as Creative Involution?: Contextualizing Deleuze and Guattari's Biophilosophy'. *Postmodern Culture* 11 (2000).
Haraway, Donna J. *Crystals, Fabrics, and Fields: Metaphors of Organicism in Twentieth-Century Developmental Biology*. New Haven: Yale University Press, 1976.
Haraway, Donna J. *Modest_Witness@Second_Millenium: FemaleMan_Meets_OncoMouse*. New York: Routledge, 1997.
Harper, Sarah and Pat Thane. 'The Consolidation of Old Age as a Phase of Life, 1945–1965'. In *Growing Old in the Twentieth Century*, edited by Margot Jefferys, 43–61. London: Routledge, 1989.
Hay, Robert and Bernard L. Strehler. 'The Limited Growth Span of Cell Strains Isolated from the Chick Embryo'. *Experimental Gerontology* 2 (1967): 123–35.
Hayflick, Leonard. 'The Limited In Vitro Lifetime of Human Diploid Cell Strains'. *Experimental Cell Research* 37 (1965): 614–36.
Hayflick, Leonard. 'Human Cells and Aging'. *Scientific American* 218 (1968): 32–7.
Hayflick, Leonard. 'The Strategy of Senescence'. *Gerontologist* 14 (1974): 37–45.
Hayflick, Leonard. 'Biological Aging is no Longer an Unsolved Problem'. *Annals of the New York Academy of Sciences* 1100 (2007): 1–13.
Hayflick, Leonard and Paul Moorehead. 'The Serial Cultivation of Human Diploid Cell Strains'. *Experimental Cell Research* 37 (1961): 614–36.
Hayles, N. Katherine. *How We Became Posthuman: Virtual Bodies*

in Cybernetics, Literature and Informatics. Chicago: University of Chicago Press, 1999.

Hayles, N. Katherine. 'Desiring Agency: Limiting Metaphors and Enabling Constraints in Dawkins and Deleuze/Guattari.' *SubStance* 30 (2001): 144–59.

Hayles, N. Katherine. *My Mother was a Computer: Digital Subjects and Literary Texts*. Chicago: University of Chicago Press, 2005.

Hayles, N. Katherine. *How We Think: Digital Media and Contemporary Technogenesis*. Chicago: University of Chicago Press, 2013.

Hegel, Georg Wilhelm Friedrich. *The Phenomenology of Spirit [1807]*. Oxford: Oxford University Press, 1977.

Heidegger, Martin. *Being and Time [1927]*. Oxford: Blackwell, 1962.

Hirshbein, Laura Davidow. '"Normal" Old Age, Senility, and the American Geriatrics Society in the 1940s'. *Journal of the History of Medicine and Allied Sciences* 55 (2000): 337–62.

Hirshbein, Laura Davidow. 'The Glandular Solution: Sex, Masculinity and Aging in the 1920s'. *Journal of the History of Sexuality* 9 (2000): 277–304.

Holliday, Robin. 'Aging is no Longer an Unsolved Problem in Biology'. *Annals of the New York Academy of Sciences* 1067 (2006): 1–9.

Holliday, Robin. *Aging: The Paradox of Life*. Dordrecht: Springer, 2007.

House of Lords. 'Judgments: The Queen on the Application of Mrs Dianne Pretty (Appellant) v Director of Public Prosecutions (Respondent) and Secretary of State for the Home Department (Interested Party)'. 29 November 2001. Available online: http://www. publications. parliament. uk/pa/ld200102/ldjudgmt/jd011129/pretty-1. htm (accessed 3 April 2014).

House of Lords. 'Ageing: Scientific Aspects'. 2005. Available online: http://www. publications. parliament. uk/pa/ld200506/ldselect/ldsctech/20/20i. pdf (accessed 17 March 2009).

House of Lords. 'Government Response to the House of Lords Science and Technology Committee Report "Ageing: Scientific Aspects"'. 2006a. Available online: http://www.publications.parliament.uk/pa/ld200506/ldselect/ldsctech/146/146we02. htm (accessed 18 April 2013).

House of Lords. 'Science and Technology Committee: Commentary on the Government Response'. 2006b. Available online: http://www.publications.parliament.uk/pa/ld200506/ldselect/ldsctech/146/14603. htm (accessed 18 April 2013).

House of Lords. 'Science and Technology Committee: Memorandum by Professor Tom Kirkwood, Specialist Adviser to Sub-Committee'. 2006c. Available online: http://www.parliament.the-stationery-office.

com/pa/ld200506/ldselect/ldsctech/146/146we07.htm (accessed 11 March 2009).
Hughes, Julian C. 'Introduction: The Heat of Mild Cognitive Impairment'. *Philosophy, Psychiatry, & Psychology* 13 (1) (2006): 1–2.
Hull, David. *Science as a Process: An Evolutionary Account of the Social and Conceptual Development of Science*. Chicago: University of Chicago Press, 1988.
Institute of Biology. *Symposia of the Institute of Biology: The Biology of Ageing*, edited by William Brunsdon Yapp and Geoffrey Howard Bourne. London: Hafner, 1957.
Jain, Dhruv. 'Capital, Crisis, Manifestos, and Finally Revolution'. *Deleuze Studies* 3 (2009): 1–7.
Jensen, Casper Bruun and Kjetil Rödje. 'Introduction'. In *Deleuzian Intersections: Science, Technology, Anthropology*, edited by Casper Bruun Jensen and Kjetil Rödje, 1–35. New York: Berghahn, 2009.
Jonas, Hans. *The Phenomenon of Life: Toward a Philosophical Biology [1966]*. Evanston: Northwestern University Press, 2001.
Katz, Stephen. *Disciplining Old Age: The Formation of Gerontological Knowledge*. Charlottesville: University Press of Virginia, 1996.
Katz, Stephen and Barbara Marshall. 'New Sex for Old: Lifestyle, Consumerism, and the Ethics of Aging Well'. *Journal of Aging Studies* 17 (2003): 3–16.
Katz, Stephen and Barbara Marshall. 'Is the Functional "Normal"? Aging, Sexuality and the Biomarking of Successful Living'. *History of the Human Sciences* 17 (2004): 53–75.
Katzman, Robert and Katherine L. Bick. 'The Rediscovery of Alzheimer's Disease during the 1960s and 1970s'. In *Treating Dementia: Do We Have a Pill for It?*, edited by Jesse F. Ballenger, Peter J. Whitehouse, Constantine G. Lyketsos, Peter V. Rabins, and Jason W. Karlawish, 104–14. Baltimore: Johns Hopkins University Press, 2009.
Kaufman, Sharon R., Janet K. Shim and Ann J. Russ. 'Revisiting the Biomedicalization of Aging: Clinical Trends and Ethical Challenges'. *Gerontologist* 44 (2004): 731–8.
Keating, Peter and Alberto Cambrosio. *Biomedical Platforms: Realigning the Normal and the Pathological in Late Twentieth-Century Medicine*. Cambridge: MIT Press, 2003.
Keown, John. *Euthanasia, Ethics and Public Policy: An Argument Against Legalisation*. Cambridge: Cambridge University Press, 2002.
Kessel, Thomas. *Phänomenologie des Lebendigen: Heideggers Kritik an den Leitbegriffen der Neuzeitlichen Biologie*. Freiburg: Alber Thesen, 2011.
Kirkwood, Thomas. 'Evolution of Ageing'. *Nature* 270 (1977): 301–4.

Kirkwood, Thomas. *Time of Our Lives: The Science of Human Aging*. Oxford: Oxford University Press, 1999.
Kirkwood, Thomas. 'The End of Age'. 2001. Available online: http://www. bbc. co. uk/radio4/reith2001/ (accessed 19 March 2013).
Kirkwood, Thomas. 'Science can Boost your Chance of Reaching a Healthy Old Age, but Don't Hold your Breath for Immortality'. *Nature* 436 (2005): 915–16.
Kirkwood, Thomas. 'Alzheimer's Disease, Mild Cognitive Impairment, and the Biology of Intrinsic Aging'. *Philosophy, Psychiatry, & Psychology* 13 (1) (2006): 79–82.
Kirkwood, Thomas and Steven Austad. 'Why Do We Age?' *Nature* 408 (2000): 233–8.
Kirkwood, Thomas and Thomas Cremer. 'Cytogerontology since 1881: A Reappraisal of August Weismann and a Review of Modern Progress'. *Human Genetics* 60 (1982): 101–21.
Kirkwood, Thomas and Robin Holliday. 'Commitment to Senescence: A Model for the Finite and Infinite Growth of Diploid and Transformed Human Fibroblasts in Culture'. *Journal of Theoretical Biology* 53 (1975): 481–96.
Kirkwood, Thomas and Robin Holliday. 'The Evolution of Ageing and Longevity'. *Proceedings of the Royal Society of London. Series B, Biological Sciences* 205 (1979): 531–46.
Kojève, Alexandre. *Introduction to the Reading of Hegel [1947]*. Ithaca: Cornell University Press, 1980.
Korschelt, Eugen. *Lebensdauer, Altern und Tod*. Jena: Fischer Verlag, 1922.
Krell, David Farrell. *Daimon Life: Heidegger and Life-Philosophy*. Bloomington: Indiana University Press, 1992.
Kroker, Arthur. *Body Drift: Butler, Hayles, Haraway*. Minneapolis: University of Minnesota Press, 2011.
Löwy, Ilana. *Between Bench and Bedside: Science, Healing, and Interleukin-2 in a Cancer Ward*. Cambridge: Harvard University Press, 1998.
Lakoff, George and Mark Johnson. *Metaphors We Live By*. Chicago: University of Chicago Press, 1980.
Landecker, Hannah. *Culturing Life: How Cells Became Technologies*. Cambridge: Harvard University Press, 2007.
Laubichler, Manfred D. and Jane Maienschein. 'Introduction'. In *From Embryology to Evo-Devo: A History of Developmental Evolution*, edited by Manfred D Laubichler and Jane Maienschein, 1–12. Cambridge, MA: MIT Press, 2007.
Laubichler, Manfred D. and Hans-Jörg Rheinberger. 'August Weismann and Theoretical Biology'. *Biological Theory* 1 (2006): 195–8.
Law, John. *Organising Modernity: Social Ordering and Social Theory*. New York: Wiley, 1993.

Law, John. *After Method: Mess in Social Science Research*. London: Routledge, 2004.
Lawlor, Leonard. 'Phenomenology and Metaphysics, and Chaos: On the Fragility of the Event in Deleuze'. In *The Cambridge Companion to Deleuze*, edited by Daniel W. Smith and Henry Somers-Hall, 103–25. Cambridge: Cambridge University Press, 2012.
Le Gros Clark, Frederick. 'The Working Life as a Measure of Ageing in Men and Animals'. In *Symposia of the Institute of Biology, No. 6: The Biology of Ageing*, edited by Geoffrey Howard Bourne and William Brunsdon Yapp, 81–90. London: Hafner, 1957.
Le Gros Clark, Frederick and Norman W. Pirie. 'Introduction'. In *Symposia of the Institute of Biology, No. 6: The Biology of Ageing*, edited by Geoffrey Howard Bourne and William Brunsdon Yapp, ix–xiv. London: Hafner, 1957.
Lefebvre, Alexandre and Melanie White. 'Introduction: Bergson, Politics, and Religion'. In *Bergson, Politics, and Religion*, edited by Alexandre Lefebvre and Melanie White, 1–21. Durham: Duke University Press, 2012.
Lemke, Thomas. *Biopolitics: An Advanced Introduction*. New York: New York University Press, 2011.
Lenoir, Timothy. *The Strategy of Life: Teleology and Mechanics in Nineteenth-Century German Biology*. Chicago: University of Chicago Press, 1989.
Lindee, Susan. *Moments of Truth in Genetic Medicine*. Baltimore: Johns Hopkins University Press, 2005.
Lock, Margaret. *Twice Dead: Organ Transplants and the Reinvention of Death*. Berkeley: University of California Press, 2002.
Lock, Margaret. *The Alzheimer Conundrum: Entanglements of Dementia and Aging*. Princeton: Princeton University Press, 2013.
Lockett, Betty A. *Aging, Politics, and Research: Setting the Federal Agenda for Research on Aging*. New York: Springer, 1983.
Longino, Helen. *The Fate of Knowledge*. Princeton: Princeton University Press, 2002.
Magerski, Christine. 'Arnold Gehlen: Modern Art as Symbol of Modern Society'. *Thesis Eleven* 111 (2012): 81–96.
Mahowald, Mary Briody. 'The President's Council on Bioethics 2002-2004: An Overview'. *Perspectives in Biology and Medicine* 48 (2005): 159–71.
Maienschein, Jane. 'Regenerative Medicine in Historical Context'. *Medicine Studies* 1 (2009): 33–40.
Marks, Harry. *The Progress of Experiment: Science and Therapeutic Reform in the United States, 1900–1990*. Cambridge: Cambridge University Press, 1997.

Marks, John. 'Molecular Biology in the Work of Deleuze and Guattari'. *Paragraph* 29 (2006): 81–97.
Martin, Moira. 'Medical Knowledge and Medical Practice: Geriatric Medicine in the 1950s'. *Social History of Medicine* 8 (1995): 443–61.
Marx, Karl. 'The Manifesto of the Communist Party [1848].' 2000. Available online: http://www.marxists. org/archive/marx/works/1848/communist-manifesto/ch01. htm#007 (accessed 30 October 2013)
Mass Observation Research Institute. 'Public Consultation on Ageing: Research into Public Attitudes Towards BBSRC and MRC-Funded Research on Ageing'. 2006. Available online: http://www.bbsrc. ac.uk/web/FILES/Meetings/ageing_mori_sri. pdf (accessed 20 March 2013).
May, Todd. 'Foucault's Relation to Phenomenology'. In *The Cambridge Companion to Foucault*, edited by Gary Gutting, 284–311. Cambridge: Cambridge University Press, 2003.
May, Todd. 'Gilles Deleuze, Difference and Science'. In *Continental Philosophy of Science*, edited by Gary Gutting, 239–58. Malden: Blackwell, 2005.
Maynard Smith, John. 'Review Lectures on Senescence: I. The Causes of Ageing'. *Proceedings of the Royal Society of London, Ser. B* 157 (1962): 115–27.
Mayr, Ernst. 'Where are We?' *Cold Spring Harbor Symposia on Quantitative Biology* 24 (1959): 1–14.
Mayr, Ernst. *The Growth of Biological Thought*. Cambridge: Harvard University Press, 1982.
McNeill, William. 'Life Beyond the Organism: Animal Being in Heidegger's Freiburg Lectures, 1929–1930'. In *Animal Others: On Ethics, Ontology and Animal Life*, edited by Peter Steeves and Tom Reagan, 197–248. Albany: SUNY Press, 1999.
Medawar, Peter. *An Unsolved Problem of Biology*. London: Lewis, 1952.
Medawar, Peter and Jean Medawar. *The Life Science: Current Ideas of Biology*. London: Wildwood House, 1977.
Medvedev, Zhores. 'Alex Comfort (1920–2000) Known and Unknown: A Personal Account'. *Experimental Gerontology* 35 (2000): 897–900.
Miller, Richard. 'Extending Life: Scientific Prospects and Political Obstacles'. *Milbank Quarterly* 80 (2002): 155–74.
Mills, Catherine. 'Playing with Law: Agamben and Derrida on Postjuridical Justice'. *South Atlantic Quarterly* 107 (2008): 15–36.
Minois, Georges. *History of Suicide: Voluntary Death in Western Culture [1995]*. Baltimore: Johns Hopkins University Press, 1999.
Mol, Anne-Marie. *The Body Multiple: Ontology in Medical Practice*. Durham: Duke University Press, 2003.
Moody, James R. 'Productive Aging and the Ideology of Old Age'. In *Productive Aging: Concepts and Challenges*, Nancy Morrow-Howell,

James Hinterlong and Michael Sherraden, 175–96. Baltimore: Johns Hopkins University Press, 2001.

Moody, James R. and Jennifer R. Sasser. *Aging: Concepts and Controversies*. Thousand Oaks: Sage, 2012.

Moore, Gregory. *Nietzsche, Biology and Metaphor*. Cambridge: Cambridge University Press, 2002.

Moreira, Tiago. 'Translation, Difference and Ontological Fluidity: Cerebral Angiography and Neurosurgical Practice (1926–45)'. *Social Studies of Science* 30 (2000): 421–46.

Moreira, Tiago. 'Truth and Hope in Drug Development and Evaluation in Alzheimer's Disease'. In *Treating Dementia: Do We Have a Pill for It?*, edited by Jesse F. Ballenger, Peter J. Whitehouse, Constantine G. Lyketsos, Peter V. Rabins and Jason W. Karlawish, 210–30. Baltimore: Johns Hopkins University Press, 2009.

Moreira, Tiago, Carl May and John Bond. 'Regulatory Objectivity in Action: Mild Cognitive Impairment and the Collective Production of Uncertainty'. *Social Studies of Science* 39 (5) (2009): 665–90.

Moreira, Tiago and Paolo Palladino. 'Between Truth and Hope: On Parkinson's Disease, Neurotransplantation and the Production of the "Self"'. *History of the Human Sciences* 18 (2005): 55–82.

Moreira, Tiago and Paolo Palladino. '"Population Laboratories" or "Laboratory Population"? Making Sense of Diversity in the Baltimore Longitudinal Study of Aging, 1965–1987'. *Studies in History and Philosophy of Biological and Biomedical Sciences* 42 (2011): 317–27.

Morelle, Rebecca. 'Scientists at Odds over Longevity'. 2006. Available online: http://news. bbc. co. uk/1/hi/sci/tech/4834128. stm. (accessed 24 August 2006).

Mullarkey, John. *Post-Continental Philosophy: An Outline*. London: Bloomsbury, 2006.

Mykytyn, Courtney E. 'Anti-aging Medicine: A Patient/practitioner Movement to Redefine Aging'. *Social Science and Medicine* 62 (2006): 643–53.

Mykytyn, Courtney E. 'Contentious Terminology and Complicated Cartography of Anti-aging Medicine'. *Biogerontology* 7 (2006): 279–85.

Mykytyn, Courtney E. 'Medicalizing the Optimal: Anti-aging Medicine and the Quandary of Intervention'. *Journal of Aging Studies* 22 (2008): 313–21.

Nelkin, Dorothy and Susan M. Lindee. *The DNA Mystique: The Gene as a Cultural Icon*. New York: W. W. Freeman, 1996.

Nietzsche, Friedrich. 'On the Uses and Disadvantages of History for Life

[1874]'. In *Untimely Meditations*, 59–123. Cambridge: Cambridge University Press, 1983.
Nietzsche, Friedrich. *Beyond Good and Evil [1886]*. Harmondsworth: Penguin, 1990.
Nietzsche, Friedrich. *On the Genealogy of Morals [1887]*. Oxford: Oxford University Press, 1996.
Nietzsche, Friedrich. 'The Eternal Recurrence and Explanatory Notes to *Thus Spake Zarathustra*'. In *The Complete Works of Friedrich Nietzsche*, 234–81, Vol. 16. New York: Macmillan, 1911.
Noys, Benjamin. *The Culture of Death*. Oxford: Berg, 2005.
Oliver, Kelly. 'Stopping the Anthropological Machine: Agamben with Heidegger and Merleau-Ponty'. *PhaenEx* 2 (2) (2007): 1–23.
Oliver, Kelly. *Technologies of Life and Death: From Cloning to Capital Punishment*. New York: Fordham University Press, 2013.
Olkowski, Dorothea. 'Deleuze's Aesthetics of Sensation'. In *The Cambridge Companion to Deleuze*, edited by Daniel W. Smith and Henry Somers-Hall, 265–85. Cambridge: Cambridge University Press, 2012.
Olshansky, Jay, Leonard Hayflick, and Bruce Carnes. 'Position Statement on Human Aging'. *Journal of Gerontology, Ser. A* 57 (2002): 292–7.
Orgel, Leslie E. 'The Maintenance of the Accuracy of Protein Synthesis and the Maintenance of the Accuracy of Protein Synthesis and its Relevance to Aging'. *Proceedings of the National Academy of Science* 49 (1963): 517–21.
Orgel, Leslie E. 'Ageing of Clones of Mammalian Cells'. *Nature* 243 (1973): 441–5.
Padui, Raoni. 'From the Facticity of Dasein to the Faticity of Nature: Naturalism, Animality, and Metontology'. *Gatherings: The Heidegger Circle Annual* 3 (2013): 50–75.
Palladino, Paolo. *Plants, Patients and the Historian: (Re)membering in the Age of Genetic Engineering*. New Brunswick: Rutgers University Press, 2002.
Palladino, Paolo. 'Discourses of Smoking, Health, and the Just Society: Yesterday, Today, and the Return of the Same?' *Social History of Medicine* 14 (2001): 313–35.
Palladino, Paolo. 'From Tragedy to Farce: On Medicine, the State and the Market'. *Journal of Historical Sociology* 20 (2007): 622–45.
Palladino, Paolo. 'Picturing the Messianic: Agamben and Titian's "The Nymph and the Shepherd".' *Theory, Culture and Society* 27 (2010): 94–109.
Palladino, Paolo. 'Overcoming the Onto-theology of the Body?: Essay Review of Giorgio Agamben's *Nudities* and Jean-Luc Nancy's *Noli Me Tangere*'. *Body & Society* 19 (2013): 123–30.

Palladino, Paolo. 'Blessed life ... : Agamben between Foucault and Deleuze'. In *Giorgio Agamben: Legal, Political and Philosophical Perspectives*, edited by Tom Frost, 207–22. London: Routledge, 2013.

Palladino, Paolo and Tiago Moreira. 'On Silence and the Constitution of the Political Community'. *Theory & Event* 9 (2006).

Park, Hyung Wook. 'Edmund Vincent Cowdry and the Making of Gerontology as a Multidisciplinary Scientific Field in the United States'. *Journal of the History of Biology* 41 (2008): 529–72.

Park, Hyung Wook. 'Senescence, Growth and Gerontology in the United States.' *Journal of the History of Biology* 45 (2012).

Patridge, Linda. 'The New Biology of Ageing'. *Philosophical Transactions of the Royal Society of London, Series B, Biological Sciences* 365 (210): 147–54.

Paul, Diane B. *Controlling Human Heredity: 1865 to the Present*. Atlantic Highlands: Humanities Press, 1995.

Pepper, John W. and Matthew D. Herron. 'Does Biology Need an Organism Concept?' *Biological Reviews* 83 (2008): 621–7.

Pickstone, John. *Ways of Knowing: A New History of Science, Technology and Medicine*. Chicago: University of Chicago Press, 2001.

Pickstone, John. 'Natural History, Analysis and Experimentation: Three Afterwords'. *History of Science* 49 (2011): 349–74.

Pollitt, Christopher. 'Joined-up Government: A Survey'. *Political Studies Review* 1 (2003): 34–49.

Popper, Karl. *The Poverty of Historicism [1957]*. London: Routledge, 2002.

Port, Cynthia. 'No Future? Aging, Temporality, History, and Reverse Chronologies'. *Occasion* (2012).

Porter, Dorothy. 'Introduction.' In *Social Medicine and Medical Sociology in the Twentieth Century*, edited by Dorothy Porter, 1–31. Amsterdam: Rodopi, 1997a.

Porter, Dorothy. 'The Decline of Social Medicine in Britain in the 1960s'. In *Social Medicine and Medical Sociology in the Twentieth Century*, edited by Dorothy Porter, 97–117. Amsterdam: Rodopi, 1997b.

Porter, Roy. 'Diseases of Civilization'. In *Companion Encyclopaedia of the History of Medicine*, edited by William Bynum and Roy Porter, 585–600. London: Routledge, 1993.

Poynton, John C. 'Alex Comfort on Self and Religion: A Case Study in the Mergence of Eastern and Western Thinking'. *Theoria* 64 (1980): 1–12.

Prins, Ad. *Aging and Expertise: Alzheimer's Disease and the Medical Professions, 1930–1990*. Amsterdam: Universiteit van Amsterdam, 1998.

Protevi, John. 'The Terri Schiavo Case: Biopolitics, Biopower, and Privacy as Singularity'. In *Deleuze and Law: Forensic Futures*, edited by Rosi Braidotti, Claire Colebrook and Patrick Hanafin, 59–72. London: Palgrave Macmillan, 2008.

Protevi, John. *Political Affect: Connecting the Social and the Somatic*. Minneapolis: University of Minnesota Press, 2009.

Protevi, John. 'Deleuze and Life'. In *The Cambridge Companion to Deleuze*, edited by Daniel W. Smith and Henry Somers-Hall, 239–64. Cambridge: Cambridge University Press, 2012.

Provine, William B. 'Ernst Mayr: Genetics and Speciation'. *Genetics* 167 (2004): 1041–6.

Pyyhtinen, Olli. 'Life, Death and Individuation: Simmel on the Problem of Life Itself'. *Theory, Culture & Society* 29 (2012): 78–100.

Rabinow, Paul. 'Beyond Ethnography: Anthropology as Nominalism'. *Cultural Anthropology* 3 (1988): 355–64.

Rabinow, Paul. *French Modern: Norms and Forms of the Social Environment*. Chicago: University of Chicago Press, 1989.

Rabinow, Paul. 'Introduction: A Vital Rationalist'. In *A Vital Rationalist: Selected Writings from Georges Canguilhem*, edited by François Delaporte, 11–22. New York: Zone Books, 1994.

Rabinow, Paul. 'Artificiality and Enlightenment'. In *Essays in the Anthropology of Reason*, 91–111. Princeton: Princeton University Press, 1996a.

Rabinow, Paul. 'Severing the Ties: Fragmentation and Dignity in Late Modernity'. In *Essays on the Anthropology of Reason*, 129–52. Princeton: Princeton University Press, 1996b.

Rabinow, Paul. 'Steps Toward a Third Culture'. In *Essays on The Anthropology of Reason*, 153–61. Princeton: Princeton University Press, 1996c.

Rabinow, Paul. *French DNA: Trouble in Purgatory*. Chicago: University of Chicago Press, 1999a.

Rabinow, Paul. 'Epochs, Presents, Events'. In *Living and Working with the New Medical Technologies: Intersections of Inquiry*, edited by Margaret Lock, Allen Young and Alberto Cambrosio, 31–46. Cambridge: Cambridge University Press, 1999b.

Rabinow, Paul. *Anthropos Today: Reflections on Modern Equipment*. Princeton: Princeton University Press, 2003.

Rabinow, Paul. *Marking Time: On the Anthropology of the Contemporary*. Princeton: Princeton University Press, 2008.

Rabinow, Paul and Nikolas Rose. 'Thoughts on the Concept of Biopower Today'. 12 October 2003. Available online: http://www.lse.ac.uk/sociology/pdf/rabinowandrose-biopowertoday03.pdf (accessed 13 February 2014).

Rabinow, Paul and Nikolas Rose. 'Biopower Today'. *BioSocieties* 1 (2006): 195–217.
Rader, Karen. *Making Mice: Standardizing Animals for American Biomedical Research, 1900–1955*. Princeton: Princeton University Press, 2004.
Rayner, Timothy. *Foucault's Heidegger: Philosophy and Transformative Experience*. London: Continuum, 2007.
Rheinberger, Hans-Jörg. *Toward a History of Epistemic Things: Synthesizing Proteins in the Test Tube*. Stanford: Stanford University Press, 1997.
Richards, Robert J. *The Romantic Conception of Life: Science and Philosophy in the Age of Goethe*. Chicago: University of Chicago Press, 2002.
Ricoeur, Paul. *The Course of Recognition*. Cambridge: Harvard University Press, 2005.
Robbins, Martin. 'The Trouble with TED Talks'. 2012. Available online: http://www.newstatesman.com/martin-robbins/2012/09/trouble-ted-talks (accessed 24 March 2016)
Rose, Nikolas. *The Psychological Complex: Psychology, Politics and Society in England, 1869-1939*. London: Routledge and Kagan Paul, 1985.
Rose, Nikolas. *Governing the Soul: The Shaping of the Private Self*. London: Routledge, 1990.
Rose, Nikolas. *Powers of Freedom: Reframing Political Thought*. Cambridge: Cambridge University Press, 1999.
Rose, Nikolas. *The Politics of Life Itself: Biomedicine, Power, and Subjectivity in the Twenty-First Century*. Princeton: Princeton University Press, 2007.
Rotman, Brian. *Signifying Nothing: The Semiotics of Zero*. Stanford: Stanford University Press, 1993.
Sarkar, Sahotra. 'The Founders of Theoretical Evolutionary Genetics: Editor's Introduction'. In *The Founders of Evolutionary Genetics: A Centenary Reappraisal*, edited by Sahotra Sarkar, 1–22. Dordrecht: Kluwer, 1992.
Sauvagnargues, Anne. *Deleuze and Art [2005]*. London: Bloomsbury, 2013.
Schumacher, Bernard N. *Death and Mortality in Contemporary Philosophy [2005]*. Cambridge: Cambridge University Press, 2011.
Seale, Clive. *Constructing Death: The Sociology of Dying and Bereavement*. Cambridge: Cambridge University Press, 1998.
Segerstråle, Ullica. *Defenders of the Truth: The Battle for Science in the Sociobiology Debate and Beyond*. Oxford: Oxford University Press, 2000.

Sengoopta, Chandak. '"Dr Steinach Coming to Make Old Young!": Sex Glands, Vasectomy and the Quest for Rejuvenation in the Roaring Twenties'. *Endeavour* 27 (2003): 122–6.

SENS Research Foundation. 'Research Advisory Board.' 2013. Available online: http://www.sens.org/about/leadership/research-advisory-board (accessed 29 March 2013).

Serres, Michel. *Le Parasite*. Paris: Hachette Littératures, 1980.

Sheldon, Joseph Harold. *Social Medicine of Old Age: Report of an Inquiry in Wolverhampton*. Oxford: Oxford University Press, 1948.

Sherratt, Yvonne. *Continental Philosophy of Social Science*. Cambridge: Cambridge University Press, 2006.

Shershow, Scott Cutler. 'The Sacred Part: Deconstruction and the Right to Die', *New Centennial Review* 12 (3) (2012): 153–85.

Shershow, Scott Cutler. *Deconstructing Dignity: A Critique of the Right-to-Die Debate*. Chicago: University of Chicago Press, 2014.

Shock, Nathan W. 'Older People and their Potentialities for Gainful Employment'. *Journal of Gerontology* 2: 93–102.

Shock, Nathan W. 'Public Health and the Aging Population'. *Public Health Reports* 76 (1961): 1023–7.

Shock, Nathan W. 'Preface'. In *Perspectives in Experimental Gerontology*, edited by Nathan W. Shock, vii–viii. Springfield: C. C. Thomas, 1966.

Simmel, Georg. 'The Metaphysics of Death [1910]'. *Theory, Culture & Society* 24 (2007): 72–7.

Singer, Peter. *Rethinking Life and Death: The Collapse of Our Traditional Ethics*. Oxford: Oxford University Press, 1994.

Smith, David W. *Essays on Deleuze*. Edinburgh: Edinburgh University Press, 2012.

Smith, Jason. '"I am sure that you are more pessimistic than I am …": An Interview with Giorgio Agamben'. *Rethinking Marxism* 16 (2004): 115–24.

Sontag, Susan. *Antonin Artaud: Selected Writings*. Berkeley: University of California Press, 1988.

Sontag, Susan. *Illness as Metaphor/AIDS and its Metaphors*. Harmondsworth: Penguin, 1991.

Stewart, John. 'The Political Economy of the British National Health Service, 1945–1975: Opportunities and Constraints?' *Medical History* 52 (2008): 453–70.

Strehler, Bernard L. *Time, Cells, and Aging*. New York: Academic Press, 1962.

Strehler, Bernard L. 'Cellular Aging'. *Annals of the New York Academy of Sciences* 138 (1967): 661–79.

Strehler, Bernard L. 'Molecular Biology of Aging'. *Naturwissenschaften* 56 (2) (1969): 57–64.
Susser, Mervyn. 'Epidemiology in the United States after World War II: The Evolution of Technique'. *Epidemiologic Reviews* 7 (1985): 147–77.
Taubes, Jacob. *Occidental Eschatology [1947]*. Stanford: Stanford University Press, 2009.
Taubes, Jacob. 'Notes on Surrealism [1966]'. In *From Cult to Culture: Fragments Toward a Critique of Historical Reason*, edited by Charlotte Elisheva Fonrobert and Amir Engel, 98–123. Stanford: Stanford University Press, 2010.
Taubes, Jacob. 'Culture and Ideology [1969]'. In *From Cult to Culture: Fragments Toward a Critique of Historical Reason*, edited by Charlotte Elisheva Fonrobert and Amir Engel, 248–67. Stanford: Stanford University Press, 2010.
Thacker, Eugene C. 'Biophilosophy for the 21st Century'. *Ctheory* (2005).
Thane, Pat. *Old Age in English History: Past Experiences, Present Issues*. Oxford: Oxford University Press, 2000.
Thévenot, Laurent. 'Governing Life by Standards: A View from Engagements'. *Social Studies of Science* 39, (5) (2009): 793–813.
Thompson, Charis. *Making Parents The Ontological Choreography of Reproductive Technologies*. Cambridge, MA: MIT Press, 2005.
Treas, Judith. 'Age in Standards and Standards for Age: Institutionalizing Chronological Age as Biographical Necessity'. In *Standards and Their Stories: How Quantifying, Classifying, and Formalizing Practices Shape Everyday Life*, edited by Martha Lampland and Susan Leigh Star, 65–87. Ithaca: Cornell University Press, 2009.
Turner, Bryan S. *Regulating Bodies: Essays in Medical Sociology*. London: Routledge, 1992.
Turner, Bryan S. *The Body & Society: Explorations in Social Theory*, 2nd edn. London: Sage, 1996.
Turner, Fred. *From Counterculture to Cyberculture: Stewart Brand, the Whole Earth Network, and the Rise of Digital Utopianism*. Chicago: University of Chicago Press, 2006.
University College London. 'Institute of Healthy Ageing, Aims of the Patridge Lab'. 2013a. Available online: http://www.ucl.ac. uk/iha/linda-partridge/aims_of_the_partridge_lab (accessed 5 March 2013).
University College London. 'Institute of Healthy Ageing, History.' 2013b. Available online: http://www.ucl.ac.uk/iha/about_us/history (accessed 5 March 2013).
University of Newcastle. 'Ageing: Our Story of Experience. 2014'. Available online: http://www.ncl.ac.uk/changingage/assets/documents/Ageing%20our%20story. pdf (accessed 25 June 2014).

Vettel, Eric J. *Biotech: The Countercultural Origins of an Industry*. Philadelphia: University of Pennsylvania Press, 2008.
Vincent, John A. 'Ageing Contested: Anti-ageing Science and the Cultural Construction of Old Age'. *Sociology* 40 (2006): 681–98.
Warner, Huber, Julie Anderson, Steven Austad, Ettore Bergamini, Dale Bredesen, Robert Butler, Bruce A. Carnes, Brian F. C. Clark, Vincent Cristofalo, John Faulkner, Leonard Guarente, David E. Harrison, Tom Kirkwood, Gordon Lithgow, George Martin, Ed Masoro, Simon Melov, Richard A. Miller, S. Jay Olshansky, Linda Partridge, Olivia Pereira-Smith, Tom Perls, Arlan Richardson, James Smith, Thomas von Zglinicki, Eugenia Wang, Jeanne Y. Wei and T. Franklin Williams. 'Science Fact and the SENS Agenda'. *EMBO Reports* 6 (2006): 1006–8.
Warnock, Mary and Elisabeth Macdonald. *Easeful Death: Is there a Case for Assisted Dying?* Oxford: Oxford University Press, 2008.
Weismann, August. *Essays Upon Heredity and Kindred Biological Problems*. Oxford: Clarendon Press, 1891a.
Weismann, August. 'The Duration of Life [1881]'. In *Essays Upon Heredity and Kindred Biological Problems*, 5–66. Oxford: Clarendon, 1891b.
Weismann, August. 'Life and Death [1883]'. In *Essays Upon Heredity and Kindred Biological Problems*, 111–61. Oxford: Clarendon Press, 1891c.
Wells, Herbert George. *The Island of Dr. Moreau [1896]*. London: Phoenix, 2004.
Williams, James. *Gilles Deleuze's Logic of Sense: A Critical Introduction and Guide*. Edinburgh: Edinburgh University Press, 2008.
Wilson, Duncan. 'Whose Body (of Opinion) is it Anyway? Historicizing Tissue Ownership and Examining "Public Pinion" in Bioethics'. In *Bioethical Issues, Sociological Perspectives (Advances in Medical Sociology, Volume 9)*, edited by Barbara Katz Rothman, Elizabeth Mitchell Armstrong and Rebecca Tiger, 9–32. Bingley: Emerald Group Publishing Limited, 2007.
Wilson, Duncan. 'Quantifying the Quiet Epidemic: Diagnosing Dementia in Late Twentieth Century Britain'. *History of the Human Sciences* (2014).
Winther, Rasmus G. 'August Weismann on Germ-plasm Variation'. *Journal of the History of Biology* 34 (2001): 517–55.
Witkowski, Jan A. 'Alexis Carrel and the Mysticism of Tissue Culture'. *Medical History* 23 (1979): 279–96.
Witkowski, Jan A. 'Cell Aging In Vitro: A Historical Perspective'. *Experimental Gerontology* 22 (1987): 231–48.
Wolstenholme, G. E. W. and Margaret P. Cameron. 'General Discussion'.

In *Ciba Foundation Colloquia on Ageing: Volume 1: General Aspects*, edited by G. E. W. Wolstenholme and Margaret P. Cameron, 238–47. Boston: Little, Brown & Company, 1955.

Yeats, William Butler. 'The Second Coming [1921]'. In *The Collected Poems of W. B. Yeats*, 210–11. London: Macmillan, 1950.

Žižek, Slavoj. *Ticklish Subject: The Absent Centre of Political Ontology*. London: Verso, 1999.

Zola, Émile. *The Beast Within [1890]*. Harmondsworth: Penguin, 2007.

INDEX

The letter *f* following an entry indicates a figure

Achenbaum, Andrew 38, 65, 110, 222 n.44
ageing 5, 20, 37–9, 55, 136, 205 *see also* longevity; senescence
  biology of 93–4
  biomedicalization of 52–4
  cinema and 216 n.9
  governance and 40–1
  literature and 68, 216 n.9
  medicine and 41–4, 46, 52 *see also* health
  research and 37, 38–9, 42–50, 55–8 *see also* biogerontoloy; gerontology
  research conferences and 76–80, 81–5
  research funding and 82–3, 85–6, 140–3
  social conventions and 86–8
  social discrimination and 50
Agamben, Giorgio 2, 25, 173–80
  death and 177–9
  Foucault, Michel and 176
  Kingdom and the Glory, The 145
  life and 178–9, 181, 195
  Open, The 3, 195, 196, 202–3
  Rabinow, Paul and 182–3
Ageing: Scientific Aspects (House of Lords) 140

Aging (Holliday, Robin) 63
Alliance for Aging Research 42–3
Alzheimer's Disease 39, 51, 224 n.38
  Mild Cognitive Impairment (MCI) and 136
  National Institute of Aging and 115–16
  University of Newcastle, Newcastle General Hospital and 134–5
Anidajr, Gil 2, 28–9, 213 n.3, 214 n.15
'Another place' (Gormley, Anthony) 9–10, 10*f*, 199–200
Anthropos Today (Rabinow, Paul) 183
Ariès, Philippe 42
Armstrong, David 88
Artaud, Antonin 163
Austad, Steven 133

Badiou, Alain 241 n.26
Baltimore Longitudinal Study of Aging 83, 107–8
Barclay, Sarah 191
Bauman, Zygmunt 41
Beast and the Sovereign, The (Derrida, Jacques) 216 n.8

Beast Within, The (Zola, Émile) 68, 70, 165
Being and Time (Heidegger, Martin) 21, 152, 214 n.9
Benjamin, Walter 195
Bergson, Henri 156–60
　Creative Evolution 156–8
　mortality and 158–9
　organism and 157–8
　reproduction and 157–8
　Two Sources of Morality and Religion 158
　Weismann, August and 157
Bergsonism (Deleuze, Gilles) 161, 166
Bernard, Claude 51, 95, 227 n.58
　Introduction to the Study of Experimental Medicine 171
Beyond the Pleasure Principle (Freud, Sigmund) 90
Beyond Therapy: Biotechnology and the Pursuit of Happiness (President's Council on Bioethics) 53
Bichat, Xavier 148, 153
　Physiological Investigations of Life and Death 4
　vitalism and 232 n.27
bio-art 183–4, 185, 241 n.20
biogerontology 4, 5–7, 10, 37–8, 48, 51–8, 62–4, 88, 118, 120, 132–9 140–3 *see also* gerontology
biology 16–23, 70, 85, 147–8, 204
　Biology of Senescence, The (Comfort, Alex) 69–72, 91
　cell and 93–4, 97, 98–9, 101–10
　evolution and 68–70, 120–3, 125–6, 167
　molecular 93–5, 111–12
　natural history and 129–30
　philosophy and 147, 152, 167–72
　populations and 81, 87, 129–31
　standardization and 100–1
Biomedical Platforms (Keating, Peter and Cambrosio, Alberto) 6, 51, 95, 96–7, 118
biomedicine 6, 37–8, 41, 43–50, 51–8, 95–7, 118, 120, 175
biopolitics 40–1, 58– 62, 176
biosociality 38, 59–60, 153
Birth of the Clinic, The (Foucault, Michel) 15, 117, 153
Blessed, Gary 134–5
Blumenberg, Hans 27, 184, 219 n.35
Bodies that Matter (Butler, Judith) 217 n.15
body, the 163–5
　caring for 5, 55, 60, 205–6 *see also* health
　embodied existence 16–22, 147
　Deleuze, Gilles and 148, 149
　Foucault, Michel and 148, 149
　Heidegger, Martin and 151
　feminist theory and 18–19
　fragmentation of 205–6
　playfulness and 186–7
　without organs 161–6, 196
Body Drift (Kroker, Arthur) 19
Boltanski, Luc (and Laurent Thévenot) 87, 139, 208–10
　On Justification 8, 32–4
Bousquet, Joë 164, 193–4, 196
Braidotti, Rosi 20–1, 150, 203, 217 n.18, 218 n.20, 230 n.7, 230 n.9
Butler, Judith 19, 217 n.17
　Bodies that Matter 217 n.15
Butler, Robert 50, 58, 61–2, 64, 114, 117–18, 177–8

National Institute of Aging and 114–15
Why Survive? Growing Old in America 114, 115

Cambrosio, Alberto (and Peter Keating) 205–6
  Biomedical Platforms 6, 51, 95, 96–7, 118
Campbell, Tim 185–6
Canguilhem, Georges 127, 225 n.3
  Bernard Claude and 51, 95
  Normal and the Pathological, The 95–6, 117
Care of the Self (Foucault, Michel) 60
Carrel, Alexis 93
Carroll, Lewis 163
Causes of Evolution, The (Haldane, J. B. S.) 69
Christianity 29, 194–5
CIBA Foundation Colloquium on Ageing 77–80, 81–2, 84
Clark, Frederick Le Gros 84–5
Comfort, Alex 73–5, 76, 83–4, 117
  Biology of Senescence, The 69–72, 91
  CIBA Foundation Colloquium on Ageing and 77, 78, 79
  gerontology and 75, 77, 86, 87–8, 112
  Joy of Sex, The 75, 89
  sexuality and 75, 89–91
Cooper, Melinda 46, 174–5, 237 n.7
Cowdry, Edmund 76, 108, 117, 227 n.58
  CIBA Foundation Colloquium on Ageing and 77–80
  Problems of Ageing 94
Creager, Angela 100

Creative Evolution (Bergson, Henri) 156–7
Cremer, Thomas 119–20, 126–7
Culturing Life (Landecker, Hannah) 100
Currier, Dianne 217 n.15
Cynicism 30, 219 n.46

Daimon Life (Krell, David Farrell) 8
Dalcq, Albert 22, 162
Darwin, Charles 129–30, 156–7, 167
  On the Origin of Species 8, 23, 156, 202
Dawkins, Richard
  life and 62, 67
  Selfish Gene, The 16–17, 59, 67, 121–3
  Weismann, August and 21
Dean, Mitchell 145, 175, 237 n.7, 239 n.51
death 42, 127, 148, 180 *see also* life; mortality
  Agamben, Giorgio and 177–8
  anxiety and 72, 154, 159
  Bergson, Henri and 158–9
  Bichat, Xavier and 148, 153
  Bousquet, Joë and 164
  Braidotti, Rosi and 20, 203
  Comfort, Alex and 74–5
  Deleuze, Gilles and 164–5
  Foucault, Michel and 4, 153–4, 176–7, 194–5
  good death 14, 44, 220 n.10, 180, 188
  Heidegger, Martin and 151–2
  Korschelt, Eugen and 151
  natural selection and 73–5
  right to die 154, 180–1, 188–93
  ritual and 188
  suicide 44, 154, 180–1, 188–91, 193–4

Taubes, Jacob and 28
Weismann, August and 72–3, 160
death drive 164
Deleuze, Gilles 2, 5, 7–8, 10, 19, 148–50, 153, 155, 160–6, 172, 202
  Agamben, Giorgio and 173–5
  Bergsonism 161, 166
  body without organs and 161–6
  Bousquet, Joë and 164, 193–4
  death and 21, 155, 164, 193–4
  desire and 149–50, 156
  Difference and Repetition 162, 166
  Foucault 4
  Foucault, Michel and 59–60, 153
  life and 64, 155, 196–7
  Logic of Sense, The 163, 166
Deleuze, Gilles (and Félix Guattari): *Thousand Plateaus, A* 165–6, 192
Department of Work and Pensions (UK) 140–3
Derrida, Jacques 7, 9, 214 n.13, 215–6 n.8, 232 n.26, 235–6 n.63
  Beast and the Sovereign, The 216 n.8
desire 149–50, 196
Difference and Repetition (Deleuze, Gilles) 162, 166
Discipline and Punish (Foucault, Michel) 18
Disciplining Old Age (Katz, Stephen) 222 n.44
disease 41, 42, 55–8 *see also* Alzheimer's Disease
  cancer 39, 102–4, 123
  costs of 139
  degenerative 63
  historical response to 128
  infectious 41, 42, 55–6

Duchamp, Marcel 183
Duration of Lifespan, Ageing and Death, The (Korschelt, Eugen) 8, 151

Essays Upon Heredity and Kindred Biological Problems (Weismann, August) 4, 126, 143, 162
evolution 6, 112–13, 143 *see also* biology; natural selection

Fate of Knowledge, The (Longino, Helen) 170
Foucault (Deleuze, Gilles) 4
Foucault, Michel 2, 7–8, 68, 147–50, 152–3, 202
  Agamben, Giorgio and 176
  Birth of the Clinic, The 15, 117, 153
  Care of the Self 60
  death and 4, 153–4, 176–7, 194–5
  discipline 130–1
  Discipline and Punish 18
  genealogy and 24–5, 65
  governance and 127–31, 176
  Hermeneutics of the Subject, The 194
  history and 24–5, 27, 31, 65
  History of Sexuality, The 40, 41, 58–9, 147, 175–6
  knowledge and power 80–6
  Order of Things, The 10, 16, 148, 153, 154, 172
  security 128–9
  sexuality and 81, 131
  sovereignty 130, 154
  vitalism and 155, 232 n.27
French DNA (Rabinow, Paul) 216 n.9
Freud, Sigmund 21, 164, 217 n.18, 234 n.46

Beyond the Pleasure Principle 90
Fries, James 43, 220 n.9

Gaudillière, Jean-Paul 100, 102
Gehlen, Arnold 29–31, 219 n. 43, 219 n. 46
genealogy 8–9, 31, 119, 204
    Foucault, Michel and 24–5, 65
    Nietzsche, Friedrich and 23
genetics 16–17, 59–60, 119–20, 131–3
    Dawkins, Richard and 62, 121–3
    Kirkwood, Thomas and 123
    Medawar, Peter and 121–2
Germinal Life (Pearson, Keith Ansell) 68
gerontology 39, 75, 77, 82–3, 86, 87–8 94, 97, 105–9, 112, 127, 134 *see also* ageing, research
    cell and 93–4, 97, 98–9, 99, 101–10, 119
    research conferences and 76–80, 81–5
    United Kingdom and 84–6, 87–8, 134–5
    United States and 82–3, 110–11
Gnosticism 25, 184–5
Gormley, Anthony: 'Another place' 10, 10*f*, 199–200
governance 40, 127–31, 179, 195 *see also* power; state policy
Governing the Soul (Rose, Nikolas) 60
Grey, Aubrey de 2, 45, 48, 137–8, 139
Griesemer, James 167, 168–9
Growing Old: Perspectives in Experimental Gerontology (Shock, Nathan) 109
Guattari, Félix and Gilles Deleuze: *Thousand Plateaus, A* 165–6, 192

Haldane, J. B. S. 90–1
Causes of Evolution, The 69
Hall, Stephen: *Merchants of Immortality* 39, 94
Hanafin, Patrick 181
Hansen, Mark 233 n.32, 234 n.50
Haraway, Donna 15, 19, 215 n.6, 217 n.15
Hardy, Thomas 68, 70
Hayflick, Leonard 39, 97–105, 108–9, 111–13, 116–17
    National Institute of Aging and 114
    National Institutes of Health and 114
Hayflick limit 93–4, 98
Hayles, Katherine 16–20, 21
health 41, 42, 138–9 *see also* biomedicine
    insurance and 42
    longevity and 42–3, 52
Hegel, G. W. F. 26
Heidegger, Martin 1–2, 7–8, 150–2
    Being and Time 21, 152, 214 n.9
    Weismann, August and 218 n.22
Hermeneutics of the Subject, The (Foucault, Michel) 194
Hirschbein, Laura 82
history 22–31, 204 *see also* geneaology
    critical 31, 120–1
    Foucault, Michel and 24–5, 27, 31, 65
    freedom and 27–8
    Nietzsche, Friedrich and 23
    Taubes, Jacob and 25–30, 200–1
    Thévenot, Laurent and 120
History of Sexuality, The

(Foucault, Michel) 40, 41, 58–9, 147, 175–6
Holliday, Robin 39, 62–3, 123–6, 137–8
  Aging 63
  cell commitment theory and 124–5
House of Lords 44, 189
  Science and Technology Committee 48, 67, 140–2
  Select Committee on Medical Ethics 189–90
Hull, David: *Science as Progress* 170

immortality 43–4, 134
  Grey, Audrey de and 45
  Holliday, Robin, Kirkwood, Thomas and cell commitment theory 124–5
  Institute for Ageing and Health 137
  Institute for the Health of the Elderly 135, 136–7
  Institute of Biology Symposium on the Biology of Ageing 76, 84–5
  Institute of Healthy Ageing 90–1
Introduction to the Study of Experimental Medicine (Bernard, Claude) 171
Island of Doctor Moreau, The (Wells, H. G.) 13, 187

Jonas, Hans: *Phenomenon of Life, The* 6
Josiah Macy Foundation 76
Joy of Sex, The (Comfort, Alex) 75

Katz, Stephen
  Disciplining Old Age 222 n.44
  sexuality and 89
Kaufman, Sharon 52–4

Keating, Peter (and Alberto Cambrosio) 205–6
  Biomedical Platforms 6, 51, 95, 96–7, 118
Kierkegaard, Søren 26–7
Kingdom and the Glory, The (Agamben, Giorgio) 145
Kirkwood, Thomas 48, 119–20, 123–6, 132–3, 137–8, 141–2
  Institute for Ageing and Health and 137
  Institute for the Health of the Elderly and 135
  Mild Cognitive Impairment (MCI) and 136
  Time of Our Lives 13–16, 20, 39, 134
Kojève, Alexandre 238 n.38
Korschelt, Eugen 73, 152
  Duration of Lifespan, Ageing and Death, The 8, 151
Krell, David Farrell: *Daimon Life* 8
Kroker, Arthur: *Body Drift* 19

Landecker, Hannah: *Culturing Life* 100
Law, John 240 n.9, 241 n.22
Lawrence, D. H. 68, 70
life 171, 203, 210
  Anidajr, Gil and 2, 28–9, 213 n.3, 214 n.15
  Bichat, Xavier and 117
  Braidotti, Rosi and 203
  Christianity and 29
  Cowdry, Edmund and 117
  Dawkins, Richard and 62, 67
  Deleuze, Gilles and 64, 173–4, 196–7
  Foucault, Michel and 117, 176–7
  genealogy and 119

governance and 40–1
quality of 181
Rabinow, Paul and 60–1, 63–4
Rose, Nikolas and 60–1, 63–4
sanctity of 181, 189, 191
Spinoza, Baruch and 174
vitalism 61, 171
life expectancy 41, 57–8, 137 *see also* longevity
Lindbergh, Charles 93
Lock, Margaret: *Twice Dead* 180
Logic of Sense, The (Deleuze, Gilles) 163, 166
longevity 2, 37–8, 39 *see also* life expectancy
  cost of 63
  demographics of 40
  health and 42–3, 52
Longino, Helen; *Fate of Knowledge, The* 170
Löwy, Ilana 100

Marking Time (Rabinow, Paul) 206–8
Marks, Harry 100
Marshall, Barbara 89
Martin, Moira 85
Marx, Karl 26–7, 150
Mayr, Ernst 69
Medawar, Peter 38, 39, 77, 79, 83–4
  CIBA Foundation Colloquium on Ageing and 77, 79
  genetics and 121–2
  Institute of Healthy Ageing and 90–1
  Unsolved Problem of Biology, An 69
Medical Research Council 85, 134
Medvedev, Zhores 87–8
Merchants of Immortality, The (Hall, Stephen) 39, 94
Milbank Memorial Fund 43

Mild Cognitive Impairment (MCI) 136 *see also* Alzheimer's Disease
Miller, Richard 50, 116
modernity 27, 60, 149–50, 153, 179, 182
  art and 183–4
molecular biology 93–4, 97, 98–9, 101–10,
  cell mortality 124–7, 134
  Hayflick, Leonard and *see* Hayflick, Leonard
  Holliday, Robin and 123–5
  Kirkwood, Thomas and 123–5, 132–3
  Strehler, Bernard and 111–12
Moorehead, Paul 102–3, 108–9
Moreira, Tiago 51, 115, 120
mortality 20, 28, 127
  age-specific mortality curve 43
  cell mortality 124–7, 134

National Cancer Institute 39
National Institute of Aging 39, 113–16, 132
National Institute of Child Health and Human Development 106, 107, 110
National Institute of Medical Research 39
National Institutes of Health 82, 104–5, 114
natural selection 6, 17, 72–4, 125, 157, 167–8
  Neo-Darwinism and 15, 17, 20, 38, 162, 167–8 *see also* Dawkins, Richard
  Weismann, August and 162, 167–8, 169
Newcastle Campus for Ageing and Vitality 228 n.29
Nietzsche, Friedrich 2, 22–3, 153, 197, 210

eternal return 197, 240 n.60
genealogy and 8–9, 23
On the Genealogy of Morals 23
Normal and the Pathological, The (Canguilhem, Georges) 95–6, 117
Noys, Benjamin 213 n.2Nuffield Trust 76, 85–6

Occidental Eschatology (Taubes, Jacob) 25–8, 200–2
Oliver, Kelly 216 n.8, 220 n.10, 229 n.30
On Justification (Boltanski, Luc and Thévenot, Laurent) 8, 32–4
On the Genealogy of Morals (Nietzsche, Friedrich) 23
On the Origin of Species (Darwin, Charles) 8, 23, 156, 202
Open, The (Agamben, Giorgio) 3, 195, 196, 202
Order of Things, The (Foucault, Michel) 10, 16, 148, 153, 154, 172
organism, the 18, 62–3, 117, 125–7, 143, 145, 166–8, 172
  Bergson, Henri and 157–8
  Deleuze, Gilles and 162–3
  Foucault, Michel and 129–30
  Kirkwood, Thomas and 125, 132–3
  Weismann, August and 126, 143

Parasite, The (Serres, Michel) 235 n.63
Park, Hyung Wook 65, 76, 222 n.44
Partridge, Linda 90, 137–8
Pearson, Keith Ansell: *Germinal Life* 68
perplication 169
Perry, Elaine 135
Perry, Robert 135
Peters, William 107
Phenomenon of Life, The (Jonas, Hans) 6
philosophy
  biology and 147, 152, 167–72
  continental/analytical division and 235 n.59
  Deleuze, Gilles and 149–50, 153, 155–6, 160–6
  Foucault, Michel and 149–50, 152–4
Physiological Investigations of Life and Death (Bichat, Xavier) 4
Pickstone, John: *Ways of Knowing* 204–5
Politics of Life Itself, The (Rose, Nikolas) 5, 60, 205
Pomerat, Charles 101
Popper, Karl: *Poverty of Historicism, The* 170
populations 81, 87, 129–31
  genetics and 69, 74, 79–80
  segmentation of 144
Poverty of Historicism, The (Popper, Karl) 170
power 145, 210 *see also* governance; state policy
  death and 154, 191–3 *see also* right to die
  embodied existence and 40, 58
  Foucault, Michel and 40, 58, 80–1, 149, 175–6
Powers of Freedom (Rose, Nikolas) 60
President's Council on Bioethics: *Beyond Therapy: Biotechnology and the Pursuit of Happiness* 53
Pretty, Diane 188–91

Problems of Ageing (Cowdry, Edmund) 94
progeneration 168
Protevi, John 169, 171, 191–3, 215 n.6, 239 n.50

quality of life 181
Quinlan, Karen Ann 178–9

Rabinow, Paul 182–4, 205, 222 n.34, 237 n.21
   Agamben, Giorgio and 182–3
   Anthropos Today 183
   art and 183
   biosociality and 58-61
   Blumenberg, Hans and 27, 184
   Deleuze, Gilles and 59, 61, 207, 216 n.9
   Foucault, Michel and 58–9, 182
   French DNA 216 n.9
   Heidegger, Martin and 150
   life and 60–1, 63–4
   Marking Time 206–8
   Rose, Nikolas and 38, 60–1, 63–4, 132, 205
replication 62–3, 167–8 *see also* reproduction; perplication; progeneration
reproduction 13–14, 63, 112–13 *see also* replication
   Bergson, Henri and 157–8
   Comfort, Alex and 75, 112
   Dawkins, Richard and 122
   Deleuze, Gilles, Guattari, Felix and 166
right to die 154, 180, 188–93
Rose, Nikolas 60, 205
   Bergson, Henri and 61
   biosociality and 58–61
   Deleuze, Gilles and 60
   Foucault, Michel and 60
   Governing the Soul 60
   life and 60–1, 63–4
   Politics of Life Itself, The 5, 60, 205
   Powers of Freedom 60
   Rabinow, Paul and 38, 60–1, 63–4, 132, 205
Roth, Martin 134–5
Rous, Peyton 103
Russ, Ann 52–4
Ruyer, Raymond 22, 162

Sabin, Albert 101, 104
Salk, Jonas 101
sanctity of life 181, 189, 191
Schiavo, Terri 191–2
Schumacher, Bernard 213 n.2, 230 n.10
Science as Progress (Hull, David) 170
Seale, Clive 188
security 128–9
Selfish Gene, The (Dawkins, Richard) 16–17, 59, 67, 121–2
senescence 71–4, 77–8, 84
SENS (Strategies for Engineered Negligible Senescence) Research Foundation 45–6, 54–6
Serres, Michel 8, 9, 211
   Parasite, The 235 n.63
sexuality 41, 75, 89–91
   Comfort, Alex and 75, 89
   Foucault, Michel and 81, 131
Shershow, Scott 229 n.30, 237 n.17, 239 n.50
Shim, Janet 52–4
Shock, Nathan
   CIBA Foundation Colloquium on Ageing and 77–80
   gerontology and 77, 82–3, 94, 99, 105–9
   Growing Old: Perspectives in

Experimental Gerontology 109
Simmel, Georg 20–1
social medicine 86, 87
social order *see also* power
   Boltanski, Luc, Laurent Thévenot and 8, 32–4, 87, 139, 208–10
   regulatory objectivity 97
   standards and conventions 86–8
sociobiology 59
sovereignty 130, 145, 154, 178, 220 n.10, 241 n.23
state policy 140, 144–5
   United Kingdom 84–6, 87–8, 134–5, 140–3
   United States 82–3, 110–11
Spinoza, Baruch 155, 173–4
statistics 102, 123, 124, 127–8
Stelarc 186–7
Stoicism 194–5, 196
Strehler, Bernard 109–10, 111
   Time, Cells and Ageing 109
suicide 44, 154, 180–1, 188–91, 193–4 surrealism 184–5
Survey Committee on the Problems of Ageing and the Care of Old People 76, 85

Taubes, Jacob 25
   fetishization and 29
   freedom and 27–8, 29
   Gnosticism and 184–5
   history and 25–30
   Occidental Eschatology 25–8, 200–2
   surrealism and 184–5
   technology, emancipation and 29–30
   theology and 25–6, 28
   time and 28

Thacker, Eugene 170–1
theology 201
   Christianity 29, 194–5
   Gnosticism 25
   Taubes, Jacob and 25–6, 28, 201
Thévenot, Laurent 97, 120
   Luc Boltanski and 87, 139, 208–10
   On Justification 8, 32–4
Thousand Plateaus, A (Deleuze, Gilles and Guattari, Félix) 165–6, 192
'Three Ages of Man, The' (Titian) 2–3, 3f, 9, 196
Ticklish Subject, The (Žižek, Slavoj) 196
Time, Cells and Ageing (Strehler, Bernard) 109
Time of Our Lives (Kirkwood, Thomas) 13–16, 20, 39, 134
Titian
   'Nymph and the Shepherd, The' 196
   'Three Ages of Man, The' 2–3, 3f, 9, 196
Tomlinson, Bernard 134–5
Treas, Judith 87, 144
Twice Dead (Lock, Margaret) 180
Two Sources of Morality and Religion (Bergson, Henri) 158

United States Public Health Service Division of Biologics 104–5
Unsolved Problem of Biology, An (Medawar, Peter) 69

Veterans Administration 110
vitalism 61, 171 *see also* life

Ways of Knowing (Pickstone, John) 204–5
Weismann, August 21–2, 67–8, 155, 167, 172
  Bergson, Henri and 157
  death and 72–3, 160
  Essays Upon Heredity and Kindred Biological Problems 4, 126, 143, 162
  evolutionary biology and 68–70, 120
  mortality and 126–7
Weismannism 69, 156

Wells, H. G.: *Island of Doctor Moreau, The* 13, 187
Why Survive? Growing Old in America (Butler, Robert) 114, 115
Woods Hole Conference on the Problems of Aging 76

zero 235 n.61
Žižek, Slavoj 20, 217 n.17
  Ticklish Subject, The 196
Zola, Émile: *Beast Within, The* 68, 70, 165

www.ingramcontent.com/pod-product-compliance
Ingram Content Group UK Ltd.
Pitfield, Milton Keynes, MK11 3LW, UK
UKHW021900220326
469204UK00008B/83